PRIMARY INSET SERIES

SCIENCE FOR CURRICULUM LEADERS

Elizabeth Clayden
and Alan Peacock

ROUTLEDGE
LONDON AND NEW YORK

First published 1994
by Routledge
2 Park Square, Milton Park, Abingdon, Oxon, OX14 4RN

Simultaneously published in the USA and Canada
by Routledge
270 Madison Ave, New York NY 10016

Transferred to Digital Printing 2006

Typeset in Century Schoolbook/Franklin Gothic
by Solidus (Bristol) Limited

British Library Cataloguing in Publication Data
A catalogue record for this book is available from the British Library

Library of Congress Cataloging in Publication Data
Clayden, Elizabeth.
Science for curriculum leaders /
Elizabeth Clayden and Alan Peacock.
p. cm. – (Primary INSET series)
1. Science–Study and teaching.
2. Science teachers–Training of.
3. Curriculum planning.
I. Peacock, Alan. II. Title.
III. Series.
Q181.C5225 1994 93–44875
372.3′5–dc20 CIP

ISBN 0–415–10390–8 (pbk)

Publisher's Note
The publisher has gone to great lengths to ensure the quality of this reprint but points out that some imperfections in the original may be apparent

Introduction to the Primary Inset Series

Several factors have combined over the last few years to create a much greater emphasis upon in-service education in schools. First and foremost, national educational reforms have ensured that there is a great deal that teachers need to know in terms of the demands of particular curriculum areas. Stress upon the core curriculum subjects of science, mathematics and English has produced programmes of study with which teachers urgently need to familiarize themselves. Second, changes in the funding arrangements for in-service work have meant that schools have been given the responsibility for planning and organizing their own in-service programmes. While in theory this is a very positive move since it allows schools to tailor their in-service to their own precisely identified needs, rather than simply having to accept what is offered by outside agencies, for primary schools, in particular, it has created problems in practice.

There is a great deal that teachers need to know...

Planning your own in-service programme to cover the full range of curriculum areas is only practicable if you either have sufficient knowledgeable specialists already on the school staff or the funds available to buy in expertise from outside.

For primary schools in neither of these happy positions, this series of packs offers a resource which will help their planning of a coherent and effective in-service programme. The packs are aimed at staff with responsibilities for curriculum leadership who may be asked to plan and run in-service sessions as part of a school's development plans. The packs do not assume any particular knowledge on the part of the curriculum leader, other than the ability to recognize and negotiate the in-service needs of their schools, and are designed to be as self-contained and as flexible to use as possible. They provide, therefore, a resource upon which curriculum leaders can draw as they see fit, secure in the knowledge that the activities in which their school staffs will be involved have been carefully planned to take them into the crucial issues in the particular areas chosen.

The following suggestions concerning the use of the packs are addressed directly to the curriculum leader who will be responsible for this use.

Planning Effective In-Service Sessions

Determining needs

The first step in the planning of an in-service programme is to make an assessment of the exact needs of the people who will be involved. This is something which can only be done by you, in collaboration with others in the school, and not by outsiders, however knowledgeable.

It is worth spending some time, in staff meetings or preliminary in-service sessions, working towards some joint statements of what these needs are. The better you can specify these needs, the more likely you are to be able to plan in-service sessions to meet them. A first step towards this might be to ask your colleagues to complete a short and fairly open-ended questionnaire asking for their views on what they need in your curriculum area. Collect these a few days before a meeting to discuss them, so that you can get a sense of the general feelings of the staff. You might present a summary of these to the staff meeting to initiate discussions on in-service planning. Do not forget to include your own views in this summary, and argue for them if you have to. As curriculum leader you are likely to have a clearer overall picture of just where you are as a staff and where you would like to go.

Aim to leave this first meeting with a list of in-service priorities to guide your immediate planning. There are two important points to note about this, however.

1. Remember priorities can, and almost certainly will, be the result of negotiation between a range of different viewpoints. It is likely that everyone asked will have slightly different views on priorities for action and the final agreed list will have to recognize all of these if it is to command the commitment it will need to be the basis of a successful programme for in-service work. This may mean that you will have to agree to include items in your list which you either do not rate as priorities or which you simply do not agree with, in exchange for the inclusion of items you think are important which most of your colleagues do not. This is what negotiation is about, and it is vital to understand this process. It is all too easy to lose a staff's commitment because they do not feel *their* views have been adequately taken account of.

2. Remember that your initial in-service plans are not set in tablets of stone. It is inevitable that, as you get under way with your plans, fresh and revised priorities will emerge which will influence your future planning. This is *not* undesirable, but in fact is the way in which real

The Christian Healthcare Leader

50 Lessons in Faith, Courage, and Calling

Matthew Bogner

Routledge
Taylor & Francis Group
A PRODUCTIVITY PRESS BOOK

Designed cover image: Shutterstock

First published 2027
by Routledge
605 Third Avenue, New York, NY 10158

and by Routledge
4 Park Square, Milton Park, Abingdon, Oxon, OX14 4RN

Routledge is an imprint of the Taylor & Francis Group, an informa business

For Product Safety Concerns and Information please contact our EU representative GPSR@taylorandfrancis.com. Taylor & Francis Verlag GmbH, Kaufingerstraße 24, 80331 München, Germany.

ISBN: 9781041294856 (hbk)
ISBN: 9781041294849 (pbk)
ISBN: 9781003769781 (ebk)

DOI: 10.4324/9781003769781

Typeset in Garamond
by Newgen Publishing UK

priorities do emerge. Needs which arise in the course of work in a particular area are usually those which are more deeply felt and important than those which are expressed as 'top of the head' reactions to initial questions.

Running sessions

Running effective in-service sessions can be a delicately balanced operation. To begin with, you, as leader, and your colleagues may have different perspectives on the enterprise which may demand some diplomacy in its handling. The participants need to feel that their existing expertise and knowledge is given adequate credit and that, as teachers with often wide experience and well-developed skills, they have something to offer to the joint enterprise of school development. 'De-skilling' them by assuming, albeit unconsciously, that they know nothing about the subjects under discussion will hinder, not help, the process of effecting beneficial change in general school practice.

Yet, as a curriculum developer, you need to feel that you are actually making progress and that things are developing, which implies that your colleagues will recognize possible shortcomings or gaps in their practice and take steps to remedy these. Holding the balance between these two pressures may require a good deal of tact and may sometimes involve your restraining yourself from making statements, especially critical ones. Successful development is always the product of negotiation. Remember, if development were simply a question of one or two people having excellent ideas, and everyone else just adopting them, there would scarcely be a need for in-service in the first place!

Another balance which needs to be achieved is that between pushing through the material you have planned and responding to the interests of the participants. Again, some tact will be required. Although it can be infuriating not to complete work you have planned because the group became side-tracked by a discussion which, to you, seemed rather a 'red herring', remember that, to others involved, it may have been a very necessary ingredient in their development. People are idiosyncratic in the way they deal with new ideas and, if there is to be any long-term effect, they have to work these ideas out in their own ways. 'Red herrings' may simply be their way of coming to terms with the ideas. They may also be, of course, nothing more than 'red herrings', in which case you will need, diplomatically, to push things on. But take care with this, and be prepared to sacrifice a little pace in your in-service programme for what may turn out to be a better quality end-product.

In general, it is a good idea to set targets for particular sessions (although you may not meet them) which all the participants are aware of. Perhaps distributing agendas might help in this. Also, requiring a group to work towards the production of a written statement over a set number of sessions can concentrate minds, especially as the agreed deadline approaches.

Monitoring and evaluating progress

You will need to keep careful track of where you have got to and, if appropriate, the decisions you have made as a staff. Although you, as curriculum leader, will need to take responsibility for this monitoring process there are several points you should bear in mind.

1. Do not keep your monitoring secret from your colleagues. There can be little less conducive to self-evaluation than the feeling that someone else is secretly evaluating you at the same time. Opening up the monitoring process will improve the quality of your development work and also help in target setting which, as explained earlier, can be a very useful way of sustaining progress.

2. The more you can involve your colleagues in evaluating their own progress the more likely you are to initiate sustained development. Take time in some in-service sessions to allow participants to think and talk about where they are and where they are going in the development of your curriculum area.

3. Everyone likes to feel they have achieved something, especially if they also feel they have worked hard. Make sure that the emphasis in your monitoring is upon achievements rather than deficiencies, and be positive and celebratory about what has been done.

Design of the Packs

The packs are organised into Units, each Unit dealing with a particular issue or area. A Unit contains five kinds of material, as described below. The Briefing Paper, Stimulus Activities, Classroom Activities and Review Session are designed so that they can be photocopied for the participants, or the curriculum leader can use or adapt their contents and photocopy whatever he or she wishes.

1. ***Leader's Guidelines*** This provides a summary for the leader of the aims and activities involved in each Unit, with suggestions for timing and guidance on running the sessions.

2. ***Briefing Paper*** These are short papers giving participants some background to the issues dealt with in the Unit. The leader's guidelines indicate how the Briefing Paper might be used, but the most common pattern will be to distribute copies of it to participants immediately following their work on the Stimulus Activities.

3. ***Stimulus Activities*** This is the introductory session for participants, consisting of activities designed to stimulate discussion and thought about particular issues raised in the Unit.

4. ***Classroom Activities*** These are activities for teachers to try out in their classrooms following their initial discussion of the issues involved. Sometimes they consist of particular work with the children and sometimes involve observing children engaged in their normal class work.

5. ***Review Session*** This is the final session for participants to review what has been observed and learnt during classroom work. The material consists of activities and discussion points.

Although the packs include some guidance on possible ways of using them in in-service sessions, it should always be borne in mind that they are first and foremost a *resource* for in-service work rather than a prescribed course. You should feel free to use them when and how it seems appropriate in your situation. In particular, you may find that you wish to vary the time you spend on particular activities in the materials, and they are designed to have this flexibility. You should always remember that the most effective in-service work is that which is tailored to meet the needs of *your* staff, rather than planned by an outsider.

Introduction to Science for Curriculum Leaders

For the first time ever, science has achieved a high status within the school curriculum for children aged 5–11. This development has encouraged teachers and researchers to raise various issues for debate, such as:

* which are the important concepts?
* what is the relative importance of the process of science?
* what understanding of science do children bring to the classroom?
* how necessary are practical, first-hand experiences?
* how can children be encouraged to ask questions, design investigations and choose appropriate ways of recording their findings?
* how should we assess a child's progress in science?

These may be new ideas for most primary school teachers – very few of whom have a specialist science background and who would consider themselves more proficient in the arts or humanities. Extensive in-service initiatives have gone some way in helping them to gain a better awareness and understanding of the subject but in a recent survey of over 900 primary teachers (Bennett *et al.* 1992) nearly 60 per cent of them said that they felt less than competent to teach science. It is also recognized that most curriculum leaders for science have a very sketchy background in this subject area.

These materials, therefore, do not assume any particular or wide-ranging knowledge of science, but are planned as a resource for curriculum leaders to use whilst facilitating their colleagues' professional development and in the preparation of school guidelines for science.

What is the Best Sequence?

Unit 1, 'Developing Children's Investigation Skills' emphasizes the importance of the processes and skills of science and sets the scene for classroom activities – it would make a useful starting point and could be followed by Unit 2, 'Organizing Workshop Activities'.

To identify needs and begin to formulate a school policy, undertaking Unit 3, 'Reviewing the Science in your School' would be the next best step. This could either be followed by Unit 4, 'Planning for Science: What Do You Need to Write Down?' or you may think that it is more appropriate to address the needs of individual teachers and thus to follow on with a selection of other Units before going back to Unit 4.

Apart from these suggestions, there is no ideal route through this material other than the one that you design taking into account the requirements of your school and colleagues.

NB All timings of activities given are 'best estimates' and should only be taken as rough guidelines. The duration of individual activities will depend very much on your group's needs and interests. Try to be as flexible as possible.

Reference

Bennett, S. N., Wragg, E. C., Carre, C. G. and Carter, D. S. G. (1992) 'A longitudinal study of primary teachers' perceived competence in, and concerns about, National Curriculum implementation', *Research Papers in Education* 7(1).

UNIT 1 DEVELOPING CHILDREN'S INVESTIGATION SKILLS

LEADER'S GUIDELINES

This Unit focuses on the crucial dimension of children's science skills, rather than their knowledge or understanding of science. It is aimed at all teachers, since the evidence of the last ten years is that many teachers, being non-specialists where science is concerned, still have difficulty conceptualizing and practising science-specific investigation skills, such as those concerned with children hypothesizing, testing and planning open-ended investigations. This Unit will enable teachers to gain experience in these areas and thus build up their confidence.

To emphasize that this skill dimension is central to all of their children's work in science, it is probably best to begin with this Unit.

Briefing Paper

It would be preferable to distribute and read this *before* undertaking the activities, in order to focus attention on using the activities to make sense of what the different skills mean, then to return to the paper during the Review Session activities.

Stimulus Activities

1. Brainstorm. This is a 'starter' to get participants to think about the concept of investigation. Emphasize that discussion is important at all stages to help teachers articulate and consolidate ideas.

2. Science skills. This might be a useful follow-up or 'homework' exercise to do next, as it requires participants to use their understanding of skills to analyse an activity. Participants will need copies of the activity sheet, entitled 'Magnets and non-magnets' (page 12) and the National Curriculum Statements of Attainment for science. Activities 1 and 2 can be done in one session and should require up to 1½ hours.

3, 4 and 5. These are planning activities, which involve coming up with designs for tests related to children's ideas (hypotheses) or problems. Helping children to identify 'testable' questions and to plan their own investigations is (according to recent pilot assessments) less well done by teachers than any other aspect of science, and therefore deserves to be given adequate time, particularly at the feedback and clarification stages. If possible, allow participants to try out the tests they have devised during the session, and encourage cross-questioning. Activities 3 and 5 could probably be carried out together in an afternoon session, with adequate time for discussion. Activity 4 probably needs such a session to itself, if teachers are going to try out their plans and discuss the outcomes fully.

You may decide that you cannot afford to devote two whole sessions to these Stimulus Activities, in which case you should use your knowledge of the participants to choose from amongst activities 3, 4 and 5 those which best match their needs. Of the three activities, number 3 is the least demanding, whilst number 4 is probably the most demanding.

Classroom Activities

Three activities are suggested:

1. Structured investigation
2. Open-ended investigation
3. Replicating measurements

You will need to discuss with participants where each of these best fits in to their programme of work for the term. The activities in themselves only take half a day each, but can obviously be repeated and refined.

(Activity 1 could well be undertaken in conjunction with Classroom Activity 1 of Unit 7, since both relate to observing group investigation work.)

Review Session

Activities 1 and 2 relate to teachers' Classroom Activities. The timing of these sessions is therefore crucial, as they need to bring together the observations and questions of different teachers who will probably be carrying out their activities at different times. One strategy might be to build a regular 'Review Session' into a staff meeting, to keep referring back to teachers' observations and reflections.

Activity 3 is a 'rounding off' activity which brings teachers back to considering their concept of Investigation by asking them to reorganize their 'mental map' of ATI into a form which helps them to see progression. It can be done as a group activity in 20–30 minutes.

BRIEFING PAPER

The material in this unit deals with three main questions:

1. What are Exploration/Investigation skills?
2. How do teachers deal with these skills in the classroom?
3. How can you go about developing your children's investigative skills?

What are Investigation Skills?

The starting point for any development of children's skills has to be, 'What do we mean by . . . ?', because the meanings of terms such as 'investigation' have not in the past been clearly defined. In particular, the notion of Investigation Skills, which is at the heart of the National Curriculum definition of Science Skills, means different things to different people, and was little used in primary science publications prior to the National Curriculum.

Recent publications such as the NCC pamphlet on 'Investigations' have begun to remedy this problem, by providing examples of activities which characterize good investigation: but teachers in any school still need to use such documents, share their understanding of what they mean, and come to a consensus – for instance, on such matters as 'what do we in this school mean by (observation) at level (3)?'. A further useful development has been the reconstruction of Attainment Target 1 in a number of ways to make it more digestible. One such reconstruction has been reproduced and used as part of Review Session 3 (see Skill Chart).

Many teachers are already noticing that, through regular reference to such charts, they are beginning to see the patterns in children's skill development. Helpful patterns can be of two kinds:

1. patterns which *categorize* skills to help memory. For example, you can see all Science Skills as helping children to:
   * *obtain* information (observing, measuring, recording, etc.)
   * *make sense of* information (interpreting, concluding, explaining, etc.)
   * *use* information (predicting, hypothesizing, testing, etc.)
2. patterns which conceptualize *level* of skill development. For example, which of the above skills taken together make up 'three-ness' – in other words, characterize a child at level three in skill terms?

Working with the charts and documents themselves can help you to internalize these patterns, which makes classroom planning and assessing so much easier.

How Do Teachers Deal with These Skills in the Classroom?

The most up-to-date evidence on the extent to which these skills are incorporated into primary teachers' science work with children can be found in reports from HMI on 'The Teaching and Learning of Science'; from the Initiatives in Primary Science Evaluation (IPSE) promoted by the ASE; and from evaluations of the pilot SAT trials (e.g. CATS). In relation to the teaching of investigation skills, all these recent reports come to the same conclusion – namely, that most work is limited to observation and description, and particularly to the observing and recording of living things. The message is that teachers commonly deal with these non-science-specific skills, in which they are relatively confident, but deal much less often with the science-specific skills in which they are less confident.

So most of the activities in this unit will be concerned with helping you to understand and practise the science-specific skills that make up the 'meat' of AT1. The ones you will focus on are:

1. Formulating testable hypotheses and designing investigations.
2. Recognizing 'fair tests' and constructing them.
3. Making measurements and assessing their reliability.
4. Generalizing from observations and deriving patterns from data.

Don't let the language put you off! A 5-year-old is formulating a testable hypothesis when she says, 'My jar holds more water than your jar', and her teacher is inviting her to do the same when she says, 'What will happen if we put some sand into the water in the jar?'. In both cases, the hypothesis can be tested by the child's doing a practical investigation: in the first case, by pouring water from one full jar into the other to see how much it holds, and in the second case, by careful observation and measurement if necessary. Likewise, a child is 'assessing reliability' if she repeats a measurement several times to see how accurate she is being in her test.

'Yes, Jim, let's see if Jar B can hold as much...'

The important word in these cases is 'testable': children will come up with many hypotheses, but they cannot all be tested by doing an investigation in the classroom.

'Fair Tests' are a central part of science and so must be reflected in the science that goes on in primary schools. The idea is important because you will get conflicting results if you do not *control* (keep the same) the things you are not concerned with, and vary only what you want to test.

Take example 10 in Stimulus Activity 2, 'Toilet paper soaks up water better than newspaper'. If you used a whole roll of toilet paper and one sheet of newspaper, you would certainly confirm the hypothesis. But if you used one sheet of toilet paper and a whole newspaper, you would get the opposite result. What matters is to compare *only* the variable you are interested in (in this case the material) and keep everything else *constant*: which means the same amount of paper, water, time allowed, etc.

Developing Children's Skills: the Reliability of Results

Another key skill which tends to be overlooked is to do with the *reliability* or *replicability* of results obtained. In other words, would you get the same result if you did it again? A good example is the experiment to do with sinking bottle tops mentioned in Stimulus Activity 2.

How many nails does it take to sink a bottle top floating on water? A child may do the test once, and come up with the answer 6. How *reliable* is that result? Were the nails placed in gently, or dropped from a great height? Was the water still or moving at the time? Many things could affect the result: the only way to get an estimate of accuracy is to do it several times and take the average, or (easier!) the median, i.e. the middle result.

Here are two sets of results for the nail experiment.

Neville's results:

1st	2nd	3rd	4th	5th
6	3	4	5	8

Charlie's results:

1st	2nd	3rd	4th	5th
5	6	6	4	4

In both cases, the median is 5 (in other words, there are two results above 5, and two below 5 in each case). However, Charlie's results show less variation – a spread of only 2 – whilst Neville's show a spread of 5 (lowest score 3, highest score 8). What do *you* deduce from this?

Activity sheet
Magnets and non-magnets

1. Look at the magnets.
   How do they feel?
   What do you think they are made of?
   Are they different colours?
   How are they the same? How are they different?
   How many ways can you sort them?
   What words could you use to describe a magnet?
2. Take a magnet for a walk around your room and find something it will stick to. Will magnets stick to ourselves? What will a magnet stick to? What will a magnet not stick to?

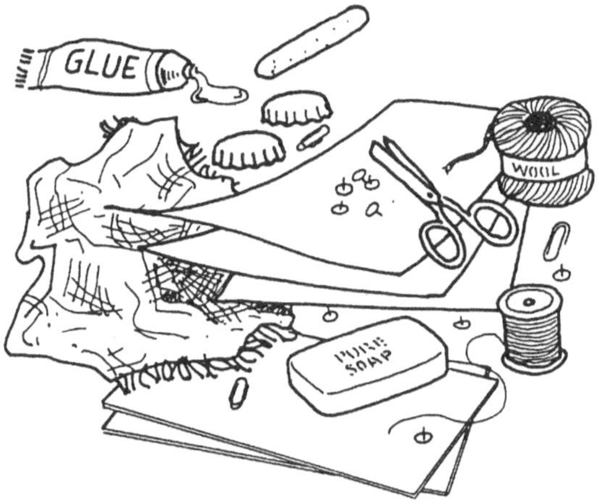

3. Look at a collection of objects. Can you guess which will stick to your magnet? Test your guess. Were you right?
   When a magnet sticks to something we say it attracts it.
   Were they big or small things?
   Are they shiny things?
   Do magnets only stick to metal things?
   Do magnets stick to all metal things?

STIMULUS ACTIVITIES

Activity 1

Brainstorm, as a group, all the possible meanings of the word 'investigation'. Make a list of all the ways in which you think your children use the word (e.g. detectives investigate crimes, plumbers investigate leaks, etc.), then decide which of these meanings are appropriate to the study of science with your children.

For example, one interpretation of the word might be 'finding things out from books in the library'. Does this fit with your idea of science skills, and is it consistent with what the National Curriculum requires?

Activity 2

Study the Activity Sheet (in the Briefing Paper) which asks children to do various things involving magnets.

Make a list of the skills involved in doing these activities, and classify the skills identified as shown below.

You should refer to the NC Statements of Attainment by number where relevant, e.g. 1.2 means AT1, level 2.

	Skills definitely involved	Skills potentially involved
Non-science-specific skills (observation, recording, etc.)		
Science-specific skills (fair testing, etc.)		

Activity 3

All the hypotheses listed below were suggested by children. Sort them out into those that are testable by children in the classroom, and those that are not.

1. My table is in the hottest part of the classroom.
2. Washing dries when the sun sucks up the water into the sun.
3. I can see further than you.
4. It takes ten nails to sink a bottle top.
5. A plant needs a lot of light because otherwise it would shrivel up.
6. The yolk of an egg is the bit a chick eats while it is growing.
7. If you boil the water it will disappear faster.
8. Everything you hit makes a sound.
9. The electricity goes clockwise round the circuit.
10. Toilet paper soaks up water better than newspaper.

Now choose *any one* of the *testable* hypotheses above, and describe how a child of the appropriate age could test it practically in the classroom.

Activity 4

Try designing fair tests around the two activities described below, and discuss your plans with colleagues. After you have finished, take it in turns to explain your plans to other participants, inviting them to 'cross-question' you. The intention is to examine if your ideas would work with children in practice.

As a group you should then try jointly to modify the activities and hold on to them as part of a 'bank' of good examples, for future reference.

1. Susan is trying to find out how quickly water drains through soil. She has collected three kinds of soil, jars, funnels, cotton wool and blotting paper, and is getting conflicting results each time she tries her experiment.
2. Ashfan is making wire loops and dipping them in washing-up liquid to blow bubbles. He has decided that making big bubbles has something to do with the size of the wire loop or the strength of the liquid, but he doesn't know which.

Activity 5

Children often make errors because they do not repeat or replicate their tests to see if their results are accurate. Discuss in your group how both of the activities listed above in Stimulus Activity 4 could be carried out so that children could replicate their results. As in Activity 4 above, you should try to come up with a suitable version of each for your 'bank' of activities.

1. Testing how far a chair can be tipped before it falls over.
2. Testing how high a seedling grows in a week.
3. Testing which paper aeroplane flies farthest.

CLASSROOM ACTIVITIES

Activity 1

Choose a session when some children in your class are undertaking a structured science investigation (i.e. an investigation where

1. the question/problem is clearly specified;
2. instructions are given as to how to carry it out;
3. materials are suggested;
4. format for recording/reporting is prescribed).

Before you observe the group, decide on a checklist you are going to use to observe the activity, which will help you to observe their Investigation Skills (AT1). You could use one of your own, or any of the lists of skills in the Skill Charts. Try first to predict what you expect to observe during their investigative work.

Plan a brief observation (10–15 minutes maximum) of a small group of children. Record all the skills that were being practised during that period, and by which children in the group.

What problems, if any, did you have with this observation?
What skills did you definitely observe being practised?
What skills were you not sure about?
Did your observations match what you predicted?

Activity 2

Choose a time when your children will be doing some *open-ended* investigative work where there is the potential for 'fair testing' (i.e. an investigation where

1. the question/problem may be loosely specified;
2. children design their own investigation;
3. children select appropriate materials;
4. children choose appropriate recording format).

Reflect first on what you learned from Activity 1 above, before planning your observation.

Plan to observe a group of children on and off during the lesson; don't be afraid to intervene and ask questions where necessary. The purpose of your observation should be:

1. To note all occasions where the need for carrying out a fair test arose.
2. To note whether or not the children recognized the need for such a fair test.
3. To look at their results and see if they were able to design and carry out a fair test.

What you are actually doing in this activity is assessing the *level* the children have reached in relation to AT1. If you have already established a record-keeping system for such formative assessment, use it to keep track of the observations you make. Keep your records, as they will be useful evidence in later Units on 'Assessment' (Unit 5) and 'Progression' (Unit 8).

Activity 3

Plan an activity for your class or for a small group, which involves them in making measurements. For example, any activity which involves measuring time (pendulums, etc.), volume (soil/water/sand investigations) or forces (stretching rubber bands, friction, etc.) would be suitable.

Do not in the first instance structure the activity by telling them what measurements to make: present it as a problem to solve. For example:

> Which rubber band stretches most easily?
> How long does it take for a pendulum to swing?
> How much water does soil absorb?

During the activity you should look for the following signs:

Do any children notice possible *sources of error* in their results?
Are they concerned about the *accuracy* of their measurements?
Are they satisfied with the *first* measurement they make?
Do they *repeat* measurements and, if so, what are their reasons for doing so?

What conclusions do you draw from your observations?

REVIEW SESSION

1. In Classroom Activity 1, you were asked to observe your children's skills during a structured science activity, and to note the problems you had in making confident observations. In the whole group, discuss your concerns and conclusions from this exercise, particularly the implications for Teacher Assessment of AT1, and make decisions about what you might collectively do next to deal with these concerns.

2. In Classroom Activity 3, you observed your class tackling activities in which replication of measurements was necessary. Refer back to what you observed, as well as to the Briefing Paper, and as a whole group suggest strategies for helping children to identify sources of error in their measurements, and for knowing when to replicate measurements and why.

3. One of the difficulties about remembering the skills in AT1 is that they are not presented in the National Curriculum document in a way which easily allows you to see the progression within a particular skill. The Skill Chart that follows is an attempt to help you 'see' progression more clearly. It reorganizes the Statements of Attainment in AT1 (levels 1–5 only, in this case) under five main areas of Investigation skills, namely:

E1 Planning
E2 Investigating
E3 Observing and Measuring
E4 Recording and Reporting
E5 Interpreting and Evaluating.

So, for example, skills in Observing/Measuring can be seen to progress from simply using the senses (level 1) through selecting and using simple instruments (level 3) to selecting and using more complex instruments accurately (level 5).

It is hoped that the above activities have helped you become familiar with AT1, so in order to let you test this, one column of the Skills Chart has been left blank. As a group, you should try to use the NC document to write in the relevant statements about Recording and Reporting skills at the appropriate levels. You might then discuss, as a group, how helpful such a chart is in your planning.

You may also find this chart useful in tackling Unit 8, 'Achieving Progression'.

Science 5–16 curriculum model: skills and contexts dimension			Exploration of science: subcomponents analysis of ATI		
Level	*E1 Planning (questions, hypotheses)*	*E2 Investigating (fair tests, handling variables, safety)*	*E3 Observing and Measuring*	*E4 Recording and Reporting*	*E5 Interpreting and Evaluating*
1.	promote ideas and seek solutions	first-hand exploration with care and safety	1a. familiar materials and events using senses		
2.	2a. how? why? what if? ideas		2b. identify simple associate differences 2c. use non-standard and standard measures		2e. associate 2 factors
3.	3a. formulate hypotheses	3c. distinguish fair and unfair test	3b. change over time 3d. select and use simple instruments and quantify to nearest division		3g. pictograms and bar charts 3h. generalize from observations
4.	4a. questions for investigation 4b. testable hypotheses 4d. plan wrt variables	4c. construct fair tests 4g. investigate safely 4f. follow instructions	4e. select and use range of instruments		4i. draw conclusions from results
5.	5a. use knowledge and skills to design investigations	5b. identify and manipulate variables	5c. select and use more complex instrument		5d. derive pattern from data

UNIT 2

ORGANIZING WORKSHOP ACTIVITIES

LEADER'S GUIDELINES

This Unit aims to help teachers develop children's investigation skills through science workshop activities. It does this by engaging participants themselves in exactly the same type of workshop activity, and by asking them to adapt the process to the children in their class.

The Unit is aimed at all teachers and can be tackled at any stage in the INSET process, although ideally it should not be done before Unit 1. Your role as leader should be to ensure that teachers are enabled to engage in the workshop activities successfully, by making appropriate time and materials available, and by focusing attention on the practicalities of achieving skills learning through the workshop. It would be helpful, before beginning this Unit, for you as a leader to refer to the Leader's Guidelines and Briefing Paper for Unit 7, 'Getting to Know Children's Ideas in Science', as these are the starting points for many workshop activities. You may decide that Unit 7 should precede this unit, but it is not essential as each informs the other.

Briefing Paper

The purpose of this Briefing Paper is to clarify and conceptualize the idea of 'workshop' activity, and the skills that such activities can develop. Ideally, the paper should be distributed and read *after* Stimulus Activities 1 and 2 have been carried out, but *before* Stimulus Activity 3.

Stimulus Activities

There are 3 activities:

1. Pattern seeking by investigating fruit
2. Identifying investigable questions
3. Hot tea

Activities 1 and 2 will require the best part of half a day. Activity 3 will require less time, probably less than an hour.

For Activities 1 and 2, you and/or the participants should collect beforehand a large selection of edible fruits. If you carry out these activities in the autumn term, you should be able to obtain a large selection of 'free fruit' from the local environment. You should supplement this with 'real' fruit from the shop, taking care to ensure that you have a good range in terms of size, colour, shape, ripeness, number of seeds, origin and familiarity. Also include some fruits which are unfamiliar; a large supermarket or Asian/Afro-Caribbean greengrocer will have a wide range and will be happy to help.

Don't worry if you yourself don't know what a particular item is; that will add to the interest of the session.

You will also need equipment for investigating the fruits: sharp knives, chopping boards, drawing materials, plastic bags for waste (if you are not going to eat the fruit!).

Divide into 3 or 4 sub-groups, and then share out the fruit so that each sub-group has more or less the same collection (this, of course, implies that your original collection will need to contain several specimens of each fruit).

You might also have a software package like Grass or Our Facts available for participants to put their results on to a database.

Activity 3: Hot tea – participants will now need copies of the Briefing Paper. You will also need to provide cups, freshly made tea and a range of insulating materials. (Do not provide thermometers at this stage, although you may wish to adapt this activity later for a more sophisticated test, in which case you will then need a way of measuring temperature.) The problem of how best to keep tea hot can be tackled in small groups or individually before comparing solutions. Leave plenty of time at the end to award the chocolate biscuit to the winner, and to discuss the practicalities of carrying out such a workshop in terms of time, resources, etc.!

In all these activities, you should help teachers to focus on:

* the *investigation skills* involved
* the *practicalities* of engaging children in such activities, from a classroom management point of view

Classroom Activities

These should be a direct adaptation of the Stimulus Activities undertaken by participants. As in Unit 1, you will need to guide some participants on the most important place in their overall plan of science work for the term/year to introduce a workshop, if they are not experienced in this way of teaching. Again, you should help teachers to focus on the *way of conducting* a workshop; simply having an activity on fruits does not in itself constitute an appropriate workshop activity. The crucial dimension is the *level of demand* made on children, in terms of Investigation skills.

Each activity will engage a group of children for at least half a day, but need not be undertaken by the whole class. For teachers less confident in investigative science, it might be appropriate to set up the workshop for a single group, and to repeat it with other groups on subsequent sessions, perhaps with a Review Session in the mean time. In this way, teachers will be able to take advantage of feedback from colleagues.

It is not essential that all teachers undertake all the classroom activities, but you should try to ensure that all teachers undertake at least one with their class.

Review Sessions

1. Report of classroom workshop activities.
2. Identifying skills developed in the workshops.
   (Ask teachers to bring to these sessions any record of their findings in the activities, such as drawings of fruits, print-outs of graphs from the database work, tables of results, etc.)
3. Brainstorming problem-solving workshop activities.

Activity 1 may be best undertaken as a series of brief, regular review sessions at which only one or two individuals report back, rather than waiting for most participants to have completed their classroom activities. These sessions could be a part of other regular staff/INSET meetings. Individual reports and reviews could usefully take up 20 minutes of such a meeting. Activities 2 and 3 are best undertaken by the whole group pooling experiences after several different types of activity have been undertaken, as they involve analysing and generating ideas. The two together might require a session of 1 hour or more.

BRIEFING PAPER

An HMI report defined the term 'workshop' as:

> *an experience of creating meaning or understanding through mental and physical action. What is provided as the basis for this action can be objects or materials to investigate or use, or evidence to examine and discuss. The outcome may be an artefact, a solution to a problem, a plan, a critique or a set of criteria; in addition to any of these, however, the product is a greater understanding of how to achieve such outcomes.*
>
> (HMI 1988)

The report went on to suggest that the essential features of workshop activity are:

1. students are active, mentally and physically;
2. the meaning of this kind of learning is conveyed through experience of it rather than through telling;
3. the students achieve an understanding, not only of the science in the problem they tackle but also of the way of learning, which they have created for themselves and so comes from within;
4. the students are involved in reflecting upon and analysing their learning experiences;
5. the product is not knowledge of a set of specific activities for children to do, but an appreciation of new kinds of learning and some of the many ways of bringing this about in children.

This Unit focuses on using workshop activity to help you teach the investigation skills of AT1. There are five types of science workshop activity, which between them incorporate all the science skills required of children in primary school. These five types of workshop are:

1. Pattern-seeking activities
2. Making Fair Comparisons
3. Hypothesis Testing
4. Pupil-initiated activity
5. Problem Solving

Unit 1 has given attention to two of these (Fair Comparison and Hypothesis Testing) by helping you to recognize these skills and design appropriate activities around them; this Unit will therefore focus on the other three types of workshop activity, by choosing suitable examples and asking you to pick out opportunities, in your own classroom, for developing such workshops.

To understand the meaning of this kind of learning fully, it is important to experience it for yourself, which is why we shall ask you to engage in an element of workshop activity yourselves. By tackling science problems, you will also be better able to diagnose children's own strengths and weaknesses in that area.

When I ask myself the question 'What do the children already know?', I do not simply mean 'How much of what I want them to know have they already acquired?'. More importantly, I need to know how their existing ideas differ from what I assume them to be, and how I should allow for this in planning their experiences.

Pattern-seeking Activity

This is the focus of Stimulus Activity 1, where you will undertake a Pattern-seeking workshop related to a collection of objects. It is always tempting to impose our own preconceptions about categories and patterns on the children, but these may not make sense to them at this stage. Thus it is more important initially to allow them to propose the categories for classification, in order that they can reveal to you their conceptualization of the area under investigation and the language they use to describe their ideas. Perhaps even more importantly, the early use of the technical 'jargon' of science can have an alienating effect on some children, making them feel that science is something apart from their normal life experience, a message completely contrary to the one we are trying to give.

An essential requirement in any workshop of this type is good data handling. The techniques used need to match the levels of attainment of the children, and so it is important to become familiar with what is appropriate at each level. The table below may help in this respect:

Key Stage 1 5–7 years	*Key Stage 2 7–11 years*	*Key Stage 3 11–14 years*	*Key Stage 4 14–16 years*
develop an understanding of the purposes of recording results and so encourage systematic recording, using appropriate methods, including block graphs and frequency charts	encourage the systematic listing and recording of data, for example, in frequency tables and bar charts	encourage the systematic recording of data using methods appropriate to the data and the purpose of the activity	encourage the systematic recording and presentation of data using, as appropriate, a full range of forms, including graphs, and mathematical relationships
encourage the interpretation of results	encourage the interpretation of data, and evaluation against the demands of the problem	encourage the interpretation and evaluation of collected data, using progressively less generalized statements against the demands of the problem, using mathematical relationships, where appropriate	involve the critical evaluation of data
	encourage the search for patterns in data	encourage the search for patterns in data and the ability to make simple predictions based on findings	encourage pattern searching in complex data and predictions requiring abstract reasoning

Children should be encouraged to handle data in various ways, to record systematically in ways which suit the data, and to interpret their own data using progressively less generalized statements. For example, a child may begin by stating that 'All fruits have skins', but will gradually move, with effective questioning, to qualify this by making such statements as 'Soft fruits have thin skins' or 'Orange skins can vary in thickness depending on the size of the fruit', and so on.

Problem Solving

Problem Solving is not confined to science; maths, language, RE and many other areas of learning all see themselves as being concerned with the solving of problems, even though the problems are of a different nature. In recent years, Problem Solving has become associated with the 'Science-and-Technology' area, and much of the thinking about problem-solving skills originated from this source, being associated with the 'Design-Make-Test-Improve' sequence. In the Technology National Curriculum, these ideas are characterized as Identifying Needs and Opportunities; Generating a Design; Planning and Making; Evaluating.

Whilst Technology is now a separate Foundation subject (and therefore not to be formally assessed), the reality is that in most primary schools much of this technology problem solving probably still goes on in close relationship to science work, for the obvious reason that the common model for science problem solving is extremely similar to that for technology, as the above examples illustrate. In fact, the Engineering Council, in its document on 'Problem solving in primary schools', specifically advocates problem-solving activities for primary children because they draw together science, craft and technology skills.

The activity on 'Hot tea' (Stimulus Activity 3) can thus be perceived as either science or technology, or science-and-technology; how would your school guidelines classify it?

The extract opposite, from a handbook for parents, summarizes the ideas above and emphasizes that it is not knowing the answers to the problems that matters; in fact, knowing answers can be a handicap, as you are more likely to steer pupils towards your solution and away from the many different solutions and processes that they might otherwise have found.

Pupil-initiated Activity

Workshops arising from pupil-initiated questions are more difficult to plan for. Some would no doubt argue that they are a luxury we cannot afford under the tight prescription of the National Curriculum. However, the evidence so far is that, at least in Key Stage 1, teachers are continuing by and large to plan their science work within the same strategies of good practice that they have used previously; that is to say, planning from themes or topics which then incorporate the Science Attainment Targets as and where appropriate. And as pointed out above, children will inevitably bring their own concerns, questions and ideas to bear on the

The Skills of PROBLEM-SOLVING

By now you may be worried that you simply don't know enough about all this to be of any help. Relax! It's not what you know that matters. It is a common feeling that, as parents or teachers, we ought to know the answers. In problem-solving, though, ***knowing the answers can be a real handicap.*** Here is an example to explain why.

The practical problem

It's breakfast time. Owen can't get the lid off a new pot of jam. He's got a real practical problem. So mum tries to do it. (Every family has a pet method! Running it under a hot tap; prizing it off with a knife; making holes in the lid; trapping it in the door jamb, . . .)

There are two snags here. First, mum's method may not work. Second, even if mum's method does work, Owen hasn't really learnt anything, except dependency. Worse, he probably assumes there is only one answer, which is obviously not true.

Recognising the problem

Here are two ways Owen might have put the problem to mum. Which would be more helpful?

This is the first essential skill. Having recognised the real problem, the knack is to express it in a way that can be solved. One of Owen's options leads him to look for causes of the problem and think about what to do next. That's a good start to solving any problem, because it sets out a clear goal.

Looking with a purpose

Owen's question made him look at the jam pot for clues. This is a vital skill in solving many problems.

Looking with a purpose, or observing, is the second skill.

Transferable skills are so much more useful than isolated facts, too. The Chinese have a proverb for it.

> ***Give me a fish and I eat for a day;***
> ***Teach me to fish and I eat for a lifetime.***

Fish are isolated facts. Fishing is a skill.

YOU DON'T NEED TO KNOW THE ANSWER

Recognising and observing are just two examples of problem-solving skills; more are given below. The important thing is that you don't *need* to know how to unstick the jam pot lid. What you *do* need to do is to get your child to look at the problem in a useful way: observe closely and think of a few possible solutions which might work. The secret is not to rush at the first idea which pops into mind. There may actually be an easier way to do it and (better still) you and your child may find it together.

Developing your problem-solving skills

If you've never tried this kind of thing, try this problem, to get the feel of it.

❝ *What is the thickness of the page of this book that you are reading?* ❞

Work at it right now. No, there's no answer at the bottom of the page! We'll come back to it on page 16. Remember:

state the problem in a useful way,

observe everything closely,

think up a few alternative ways of solving the problem.

topic being covered, which have somehow to be dealt with. The Science Working Group's final report expressed it as follows:

> *In their early experiences of the world, children develop ideas which enable them to make sense of things which happen around them. A child brings these informal ideas to the classroom, and the aim of science education is to adapt or modify these ideas to give them more explanatory power. Viewed from this perspective, it is important that we should take a child's initial ideas seriously, so as to ensure that any change or development of these ideas, and the supporting evidence for them, makes sense and, in this way, become 'owned' by the child.*
> (DES 1988, para. 2.9)

What better way to encourage 'ownership' than by following up an enquiry instigated by children? This does not mean having thirty different science investigations going on simultaneously in your classroom; you could never resource or supervise such mayhem! It means capitalizing on the correct moment at which to turn a child's question into a class or group activity as a logical development of your exploration; and consequently allowing opportunities for this to happen in your original scheme of work.

This might seem to involve taking a planning risk which requires a lot of confidence; but when you begin to observe your class working at investigations, you will notice many examples of them making statements, suggesting explanations and theories, asking questions of one another (or just into the air!), wondering and thinking aloud . . . and from this mass of child-initiated talk there is bound to emerge the trigger or starting point that leads in a direction you can profitably go next.

References

Department of Education and Science (DES) (1988) *Science for Ages 5–16*, London: HMSO.

HMI (1988) 'Implications of the National Curriculum in Primary Science for initial teacher training', Paper presented at HMI Invitation Conference on Science in the National Curriculum, Blackpool (Nov.).

Peacock, A. (1989) 'The skills of problem solving', in *SUCCESS! The Essential Parent Guide*, Basingstoke: Macmillan Education, pp. 13–16.

STIMULUS ACTIVITIES

Activity 1

Using the large collection of fruits that you should now have in front of you, sort your fruits into categories according to any criteria you consider to be appropriate. Resist the temptation to be 'clever' and use 'scientific' categories straight away; and don't let the botany 'expert' in your group tell you what the categories should be! When you have sorted the fruits in several different ways, compare your categories with those of other groups. You will probably find that some of their categories were the same as yours, whilst some were different.

Now choose four or five of the categories, and assign all the fruits to them as appropriate. For example, if one category is *size*, you may decide for each fruit whether it is small, medium or large; or you may decide to do it by measuring (up to 5 cm, up to 10 cm, up to 20 cm and so on).

To do this, you will need to keep some form of record or it will be impossible to see any patterns in your data. You can use drawings, tables, graphs, or put the results on to a database which will enable you to do some analysis.

All the above is the lead-up to pattern seeking. You can now, as a sub-group, start to *look for connections* between the data on the fruits; e.g. is there any connection

* between *size* and *colour*?
* between *taste* and *number of seeds*?
* between *hardness* and *skin type*?

If so, what is the connection, or the 'pattern'?

At the end of the session, share your ideas with other sub-groups. Did you all come up with something different? Good! Why do you think this was? What did you learn from this? Is there a 'right' and 'wrong' way? Were all your ideas 'right'?

Activity 2

The activity on fruit must have raised several questions which the pattern-seeking activity itself didn't answer. As a group, write down a 'brainstorm' list of all the questions raised, the answers to which you are genuinely interested in finding out.

Having done this, go through the list and decide which of these can be investigated practically, deleting from the list those which cannot.

Here are two examples to give you the idea.

Q1: What is the best way to preserve a (banana/mango/cucumber)?
Q2: What makes an apple turn brown when you cut it open?

The first question *can* be tackled by trying out a range of ways of preserving whatever fruit you are interested in: putting it in the fridge, wrapping it in foil, etc. The second question could perhaps be answered by library research, but *not* by any experiment which is feasible in a primary classroom. However, the question can be adapted into a testable form: for example, you might ask, '*How can you prevent* an apple

turning brown when it is cut?'. Such a strategy is often important with children, to turn their real but impractical queries into investigable questions.

Now choose one question from your 'testable' list and design a way of investigating it. Then try out your idea and see if it works. Was it in fact feasible? If not, what were the problems? What have you learned about pursuing child-initiated questions?

Activity 3

Here is a simple problem which you can tackle whilst you have your cup of tea!

The problem is this: of all the teachers in your group, who can keep their tea hot longest, and how did they do it?

Refer to the problem-solving sequence set out on p. 24 (Design, Make, Test, Improve) of the Briefing Paper before you begin.

CLASSROOM ACTIVITIES

You should now modify each of the three workshop activities you undertook in the Stimulus sessions to suit your class of children and the topic you are currently covering, then try them out, either with a group at a time or with the whole class.

For example, Stimulus Activity 1 on fruits might be adapted to a pattern-seeking activity related to another appropriate collection your children are making (leaves; rocks; utensils; fabrics; metals; paper; liquids). Alternatively, you could use an activity which simply involves collecting data about two related variables, such as the change in length of a rubber band when it is stretched using a forcemeter or weights.

Also, the way of recording and tabulating would have to be adapted to their known level of attainment in this field. You might refer to the section on 'Pattern-seeking activity' in the Briefing Paper, or to the SEAC guide to 'Graph work in school science' referred to in the Briefing Paper before finalizing your plans.

Carry out the workshop with your class, but *before* you do so, make a few notes about what you *expect* to happen. During the activity, try to avoid suggesting explanations or answers to children's questions; try to return their question with another which necessitates further investigation. For example:

CHILD: Are all metals shiny?
TEACHER: Can you find any in your collection which aren't?

Don't try to observe everything during the activity: note down specifically all their statements or questions which refer to possible patterns within the data they are collecting. Bring all these notes about their suggested patterns with you to the review session, along with examples of whatever work or display material the children produced.

If you can't find time to try out all the three different workshop activities with your class, or if they simply don't fit in to your scheme of work for the term, don't worry; but ask yourself, if I am not fitting in any of these workshop activities this term, how am I making sure that the relevant Investigation skills are being covered? You should certainly make sure that at least one type of workshop is adequately dealt with, and that you have evidence to bring to the group discussion.

REVIEW SESSION

1. Individually describe the workshop activities you have carried out in your classroom, focusing on:

* what the children actually did
* what you think the children learned, in terms of investigation skills
* your role as teacher during the workshop
* the problems you and the children encountered

Other members of the group can then suggest ways in which the problems might be overcome.

2. First remind yourselves of the investigation skills appropriate to the children in the age range which you teach (for example, by looking back over your review session in Unit 1 if appropriate).

Then as a group discuss which type of workshop activity was best suited to developing each of these skills in children. You might come up with a chart something like this:

	Pattern seeking	Problem solving	Pupil initiated
Observing	✓	✓	✓✓
Comparing	✓✓		
Measuring			
Predicting		✓✓	
Fair testing		✓✓	✓
Etc.			
Etc.			

Where ✓✓ indicates 'most appropriate', ✓ indicates 'sometimes appropriate', and so on.

Did you find any essential skills which were not adequately dealt with by these activities? How, then, do you ensure that they are developed by children?

3. Stimulus Activity 3 involved finding a practical solution to an everyday problem (keeping tea warm) and learning some science skills in the process.

Brainstorm a list of other everyday problems encountered by the children in your class, or by the teachers in your school, which can be tackled by the same problem-solving workshop activity. Try to focus on problems which are *genuine* problems, and which can be tackled using *everyday* materials available in the classroom.

For example, to start you off, you may consider problems such as:

* drying wet clothing
* keeping milk cool
* removing stains
* watering plants and feeding fish in the holidays
* and many others!

References

Taylor, R. M. and Swatton, P. (undated) 'Graph work in school science: a booklet for teachers', Assessment Matters no. 1, London: Schools Examinations and Assessment Council.

UNIT 3

REVIEWING THE SCIENCE IN YOUR SCHOOL

LEADER'S GUIDELINES

This Unit is intended to encourage discussion about the planning, teaching and assessment of science in your school. As a review of present policy and practice, it is seen as the first step in identifying the needs of both the school and the staff. You may wish to follow this Unit with Unit 4, 'Planning for Science: What Do You Need to Write Down?', or to delay this work and to focus instead on some of the aspects of science that are addressed by other Units and that are identified as 'areas of concern' during the Review Session.

Briefing Paper

You should distribute this (and the questionnaire that forms the focal point for Stimulus Activity 1) before the first session, so that the course participants are able to use it to help prepare for this session.

Stimulus Activities

There are four activities which initiate the review and encourage participants to talk about their science planning, teaching and assessing. They are then asked to focus on those aspects of planning in the school that they think require attention and to identify the tasks that need to be carried out in the form of an action plan.

1. Reviewing Policy and Practice. The questionnaire should be handed out to the teachers before the first session. They should be asked to read it and to start to answer the questions.
2. This is a discussion of the questionnaire – you should allow up to 1½ hours.
3. The first part of the action plan. If the group of teachers is large enough it can be split into four, each smaller group dealing with one of the four headings. Allow 1½ hours for discussion and plenary sessions. Alternatively, the whole group can have a discussion using all four headings, one at a time, followed by a summing up. For this you will need **two** 1½-hour sessions.
4. This is the second part of the action plan and is a discussion of the roles staff play in curriculum development. Allow about ¾ hour for this.

Classroom Activity

There is only one Classroom Activity, and you should allow up to three weeks for this.

1. Observing science work in the classroom. You may wish to allow

participants to choose two of the three weeks for an analysis of the science taking place in their own classrooms. Participants will need to be aware that the discussion in the Review Session will require the information that they are gathering during this time.

Review Session You should allow up to one half day for this session. Participants will require this amount of time in order to share and discuss the observations that they have made in their own classrooms, to identify their needs and aims for the future, and to match these with a further programme of work. They will need to refer again to the questionnaire in Stimulus Activity 1, and in order to decide which other Units they wish to undertake they will need to look at the complete pack of 'Science for Curriculum Leaders'.

BRIEFING PAPER

Science in Primary Schools

Recent reports from HMI suggest that the amount and quality of science now taught in primary schools has increased since the earlier surveys of the late 1970s and the 1980s. Many of you will remember that despite extensive national initiatives – such as the Nuffield Junior Science Project and the Schools Council Science 5–13 – the National Primary Survey of 1978 had little that was good to say about this area of the curriculum! It suggested that teachers lacked a working knowledge of science; that schools lacked effective programmes and equipment for the teaching of science; that there was no proper coverage of key scientific notions; that the teaching of processes and skills was often superficial; and that 'the work in observational and experimental science was less well matched to children's capabilities than work in any other area of the curriculum' (DES 1978).

The Department of Education and Science issued its first Statement of Policy in 1985 – 'Science 5–16' – which stated that 'All pupils should be properly introduced to science in the primary school' (DES 1985). This document can now be seen as a forerunner to the National Curriculum. It identified the appropriate content for science teaching, and stressed the importance of the methods of science; that is to say, the relevant skills and processes. Schools were required to appoint curriculum leaders for science and to formulate policies for teaching it, taking account of ten

principles including breadth, balance, relevance, continuity, progression and links across the curriculum.

The National Curriculum has, of course, taken these ideas further and has identified Programmes of Study for each of the Key Stages and Statements of Attainment to use when assessing the achievements of each child. The skills and processes (the methods of science) are found within Attainment Target 1, and it is suggested that for all science-based activities there should be a balance between the ideas in this AT and those that form part of the other ATs – the 'knowledge and understanding'.

Effective implementation and management of the National Curriculum for science in your school requires you to think about and plan

* the time that will be required for science teaching and learning, as part of the core curriculum
* the balance between the teaching and learning of knowledge and understanding and of scientific skills
* continuity and progression of science teaching and learning within each classroom and between the different classes or year groups
* the ways in which children's progress will be assessed, and their achievements recorded

The development of any curriculum area is best initiated with a review of present practice within the school. It is also important that each member of staff is aware of the role that he or she is to play during the planning, management and teaching of this area of the curriculum. Although such questions can be resolved solely by the headteacher, they are likely to be more useful starting points for development and change if all the staff are able to contribute to any review or decision making that needs to take place. Most of the activities within this Unit have been planned with this idea in mind.

References

Department of Education and Science (DES) (1978) *Primary Education in England – A Survey by HMI*, London: HMSO.

—— (1985) *Science 5–16: A Statement of Policy*, London: HMSO.

—— (1989) *Science in the National Curriculum*, London: HMSO.

—— (1991) *Science in the National Curriculum*, London: HMSO.

Nuffield Junior Science Project (1967) London: Collins.

Schools Council (1972) *Science 5–13*, London: Macdonald Educational.

STIMULUS ACTIVITIES

Activity 1

A review and evaluation of present practice in the school is one of the first stages of curriculum development. It is a way of identifying needs, both of the school and individual members of staff. Priorities for future initiatives and for INSET can then be established. Although such a review can be carried out solely by the headteacher, it is likely to be more useful as a starting point for development and change if all the staff are able to contribute. The following list of questions has been designed to help you with your review of the science that takes place in your school and classroom. It will be used during this Unit, and you will find it useful to think about it before the first session takes place.

Questions to use when reviewing the planning, teaching and assessment of science in your school.

Planning

1. Do we have a scheme of work for science in the school, or do individuals plan their own?
2. How do we ensure that we take account of
   * the Programmes of Study in the National Curriculum
   * HMI/Adviser's reports
   * LEA guidelines
   * parental expectations
   * children's progression and achievements?
3. Do we link science with other curriculum areas?
4. How do we ensure that there is a progression in the development of skills and of knowledge and understanding throughout each Key Stage?
5. Do we all know what resources we have for science and where they are kept?

Teaching

1. Do we ensure that the science activities that take place in our classrooms
   * show a balance of all the components of AT1
   * show a balance between AT1 and the other ATs?
2. If there is a school plan for science, does it provide a platform for the development of schemes of work in my classroom?
3. Do we
   * set clear objectives for scientific investigations
   * use familiar and interesting starting points
   * involve the children in aspects of planning
   * use open-ended questions to guide their thinking
   * give them time to discuss and make predictions
   * provide them with opportunities to interpret their findings?
4. Can we easily find and collect the resources that we require?

Assessment

1. Do we make opportunities to assess skills other than those of observation and recording? That is, do we assess how the children carry out an investigation as well as what they produce at the end of an activity?
2. Do we make opportunities to assess children's knowledge and understanding?
3. Do we involve the children in assessment procedures?
4. Do we use what we find out from assessment in a formative way – to help plan what the children do next?

Activity 2

Read and discuss the questions that you were asked to consider in Activity 1. Add any others that you think may help your review of the planning, teaching and assessment of science in your school.

Activity 3

In the light of your discussion during Activity 2 and using the headings set out below:

* decide what changes need to be made in the ways in which science is planned in your school
* identify the tasks that need to be carried out and plan a realistic time scale for their completion

1. taking account of the expectations of others;
2. linking science to other curriculum areas;
3. ensuring progression;
4. identifying resources required.

Activity 4

Discuss the existing roles of the headteacher, deputy head, science coordinator and individual teachers in the planning and teaching of science within the school. You may also like to include support agencies, such as advisory teachers who regularly visit the school.

Do these existing roles match the tasks that were identified previously? If not, how do they need to be changed?

CLASSROOM ACTIVITIES

Observe and describe what the children in your class do in science during a two-week period.

Pay particular attention to:

your planning

* its relationship to any school guidelines
* its relationship to work that the children have done before
* its relationship to other curriculum areas
* the starting points

your teaching

* the ways in which the children carried out and recorded their work
* how did you guide their thinking?
* what were they learning about?
* did they have all the resources that they required?

your assessment

* what did you look at in order to gain an understanding of the children's achievements?
* how did you record the findings of your assessment?

Bring your observations for discussion at the Review Session.

REVIEW SESSION

Compare what each of you has found out about the science in your classrooms as experienced by children.

Use the questionnaire that formed part of Stimulus Activity 1 to help you to discuss your findings, and to

* decide what changes need to be made in your classroom and throughout the school
* identify your needs for INSET and match these with a sequence of Units from *Science for Curriculum Leaders* that you wish to undertake
* identify any other tasks which need to be carried out, linking them with those that you talked about in the first session of this Unit, and plan a realistic time scale for all this work

UNIT 4

PLANNING FOR SCIENCE: WHAT DO YOU NEED TO WRITE DOWN?

LEADER'S GUIDELINES

This Unit is designed to help teachers to identify the documentation they need related to their planning and teaching of science, and to agree on a plan of action to write the documents they require.

The most important part of this process is the relevant discussion that can be generated by the Briefing Paper and by undertaking the various activities.

It is important that this Unit is not attempted until after Unit 3, 'Reviewing the Science in your School'; the accompanying Briefing Paper could also be of help here.

Briefing Paper This should be distributed before the first session, and teachers should be encouraged to refer to the current National Curriculum document for science and the non-statutory guidance material for Science (NCC 1989) whilst they read the Briefing Paper.

Stimulus Activities There are two activities, and they both concern reviewing existing documents for the teaching of science. These activities are for those who already have some documentation, which they should bring to the session. The time needed for this session will depend on the number of documents available.

Those teachers who do not have any existing documents should read the Briefing Paper and discuss what they require.

1. Evaluating documents currently in place
2. Criteria checklist for planning documentation

Classroom Activities There is only one Classroom Activity: 'Planning for Science'. You should allow teachers at least two weeks in which to make notes about the planning that they do for science work during a two-week period. It is important to emphasize that they should not do anything different from their usual practice, as an analysis of current practice is the most fruitful way of providing ideas for the future.

Review Session The aim of this session is to decide on an action plan for the future. If possible, you should try to obtain documents from other schools, as these can often provide ideas.

1. Discussion of current science planning and the use made of documents.
2. Deciding what documents need to be revised and what need to be written.
3. Deciding on an action plan for the production of relevant and useful documents.

The organization of this session will depend on the number of participants. You will have to decide whether they work in groups of four or five for '1' and '2' – reporting back to the whole group during a plenary session – and then everybody tackling '3' together, or whether – particularly if it is not a large group of participants – there should be a more general discussion. In any case, you should allow 1½ hours for this session. Someone – you? – must act as a chairperson. (Some tact may be required here; it is important that no one becomes disheartened, especially those who find science difficult.) You will need to guide the discussion so that appropriate decisions are made, ensuring that everyone understands their role in the plan.

References

National Curriculum Council (NCC) (1989) *Science: Non-statutory Guidance*, London: HMSO.

BRIEFING PAPER

A Rationale for 'Writing Things Down'

It has become increasingly common to find a range of documents within a school that have been written by some or all of the staff and that describe policies and give guidelines for teaching the various areas of the curriculum. Indeed, schools are now required to develop schemes of work which describe, in broad terms, what will be included in the work of each classroom. Various reports and surveys, from HMI and others, stress the desirability of policies and schemes, and indicate that the implementation of the curriculum within a school benefits from the existence of written guidelines.

The publication of a glossy booklet is sometimes seen as the major goal of curriculum development, whereas this exercise of putting things on paper is only part of the process. Teachers' attitudes and styles of work within the classroom, ways of recording children's progress, identification of necessary resources, and ways of communicating with parents are all important aspects of curriculum development. The results of the ILEA Junior School Project suggest that where staff are involved in the development of curriculum guidelines there is more likely to be consistency in guideline usage (ILEA 1986: 26). Consequently, a working document, developed by the staff and subject to regular evaluation, is probably a more appropriate way of expressing a policy than something which looks like 'tablets of stone' and which sits, unused, in the headteacher's office awaiting a visit from the local inspector.

What Documents Do We Need and Who is Going to Use Them?

The non-statutory guidance material for science from the National Curriculum Council in 1989 says that

In order to provide a coherent structure for the science curriculum in a

school, it is essential that teachers have a shared vision of the purposes of their teaching. The development of the scheme of work, by teachers working together, is a mechanism whereby a coordinated approach can be achieved. The scheme of work is an important part of the school's curriculum policy for science. (NCC 1989)

It is suggested that each school should develop:

1. a *policy* for the teaching of science which will contain
   * broad aims about the nature and scope of the science to be taught. In other words, *how* the school intends that science should be taught and *what* topics are going to be included.
2. a *scheme of work* for science which will
   * reflect the ethos of the school
   * expand on the school's policy for science
   * describe the work planned in detail over a specific period, for each class or group
   * use the programmes of study and the Attainment Targets as a checklist, ensuring coverage of what they contain whilst at the same time leaving room for flexibility
   * avoid unnecessary repetition and ensure progression and continuity
   * contain the component *areas of study*.
3. the *areas of study*, which can be
   * *topics* like 'the river', 'celebrations' or a story, and which usually have a cross-curricular approach so that the science part is identified and has a varying emphasis, according to the nature of the topic, or
   * *themes* like 'ourselves' and 'air and flight', which are science based. Several 'narrow' themes can be found together as part of a 'broad' theme; for example, 'ourselves' could be part of the broad theme of 'living things', or
   * *units of science work*, which may be based on a resource such as gardening or cooking, or which may be quite separate from what is currently being taught in the class, especially if the more general topic contains very few opportunities for science activities. 'Electricity' is often taught in this way.

Each area of study needs to be based on one or more Attainment Targets, related to a range of levels of attainment and focused on the appropriate Statements of Attainment.

You might like to add to this framework and include information about

* the school's resources for science, and where they are
* some different approaches to organizing science activities in the classroom
* some ideas for starting points for science
* ideas about safety in the classroom
* strategies for children with special needs

References

Department of Education and Science (DES) (1989) *Science in the National Curriculum*, London: HMSO.
—— (1991) *Science in the National Curriculum*, London: HMSO.
Inner London Education Authority (ILEA) (1986) *The Junior School Project*, London: ILEA.
National Curriculum Council (NCC) (1989) *Science: Non-statutory Guidance*, York: NCC.

STIMULUS ACTIVITIES

Activity 1

It is important that before you write any new policies or schemes for the planning and management of science in your school, you identify the documents that you have already, and evaluate them. You will have been asked to bring to this session all the documentation that you have – whether you use it or not!

Sort this material according to whether it is written for

* the whole school
* the year group
* an individual class

Describe how you use each of these documents (if they are used).

Do any of them include a policy for the teaching of science in your school?

Do any of them include a scheme of work containing areas of study? (You may want to refer to the section of the Briefing Paper which describes these.)

Activity 2

Below is a checklist of criteria for the content and use of school science guidelines. Discuss these and add any other criteria you think are important for you and your school.

How do any documents that you have at the moment match these criteria? Evaluate each in turn.

Checklist to use when planning documentation for science

1. Does it indicate the views held by the teachers of the school about science in the primary classroom?
2. Does it start with general statements, moving towards the more particular?
3. Will it encourage balanced science teaching?
4. Will it encourage the children to develop a progressive understanding of science skills and concepts?
5. Does it address the issue of continuity?
6. Does it give practical help for teachers when planning science activities in their classrooms?
7. Is it easily understandable; is it user-friendly?

You may like to add other points

CLASSROOM ACTIVITY

Make notes about the planning that you do for your science work with children during a two-week period.

How do you actually plan?

What do you refer to?

How do you ensure that there is

* a balance between the various aspects of AT1, and
* a balance between AT1 and the other ATs?

Are you able to refer to what the children have done before

* in your class
* in a previous class?

How will you pass on information to the next class teacher about what the children have done and what they have achieved?

If your school has any written plans for science then

* which document(s) have you found to be of most use during this two-week period? why?
* which document(s) was of least use? why?

Bring your notes to the Review Session.

REVIEW SESSION

1. Share and discuss what you have found out about

* the planning that you each carry out for teaching science in your classrooms
* the references that you each make to any documents that the school has already
* your needs for documents at school, year and class level

2. Refer to the evaluations that you made previously of any documents that you have already, and decide

* which documents should be revised
* which new ones should be written

3. Decide on a plan of action that will result in the production of relevant, useful documents for you as class teachers and for the school as a whole. You may like to refer back to

* the checklist of criteria in Stimulus Activity 2; are there any others that you now think ought to be included?
* the ideas from the documents of other schools

For each document decide

* who is going to use it?
* when and how is it going to be used?

UNIT 5

ASSESSING CHILDREN'S LEARNING IN SCIENCE

LEADER'S GUIDELINES

The aim of this Unit is to help teachers to develop the skills that are required to assess children's learning and understanding of science. Planning for assessment is another underlying theme, and its importance can be emphasized by linking this Unit with Unit 4, 'Planning for Science: What Do You Need to Write Down?'.

Briefing Paper

This should be distributed before the first session, so that course participants can begin to focus on the ideas within the Unit. It can then be referred to during both the first session and the review.

Stimulus Activities

There are four activities. Up to 1½ hours should be allowed for each activity, although Activities 1 and 2 could be extended by the provision of other examples of children's work.

1 and **2.** Assessing children's written work
3 and **4.** Planning for assessment

Teachers will need the National Curriculum science documents (GB, DES 1989 and 1991) to enable them to carry out these tasks.

Instead of undertaking these activities in sequence, another strategy would be to carry out the first two Stimulus Activities and to follow with Classroom Activities 1, 2 and 3 before going back to Stimulus Activities 3 and 4.

Classroom Activities

1. Assessing children's work
2. Observation and assessment
3. Identifying assessment opportunities
4. Using Statements of Attainment

Again, course participants are encouraged to take part in assessment and in planning for assessment – but this time within their own classrooms. You may like to distribute the activities amongst the group so that each person undertakes, say, two during a fortnight. Participants will have had different experiences with assessment, and this should be taken into account when you are distributing the activities. You will also need to bear in mind the organization of the Review Session, where the participants are asked to discuss their work with a colleague who has been engaged in a similar activity.

Review Session

1. Interpreting Statements of Attainment
2. Sharing assessments
3. Developing assessment

The most effective way to organize this session is to divide the group into smaller groups or pairs, each of which then undertakes one of the activities. Then, depending on the number of people present, individual groups can either report back to the whole group or they can extend their discussions with another of the small groups before taking part in a general reporting session. The second of these patterns of procedure is implicit in the description of Activity 1, and is particularly relevant to the focus of this activity.

You will need to divide up the time available in an appropriate way, leaving enough time for a general discussion at the end.

BRIEFING PAPER

Before the advent of the National Curriculum, the word 'assessment' was rarely used when we talked about what went on in a primary classroom – and yet this does not mean that it didn't happen! Teachers, then as now, were continually noting what individual children achieved and what they found difficult, in order to decide on their next most appropriate task or experience. The information that was passed on to parents was also based on the teacher's assessment of each child. During the last few years assessment has become recognized as a well-defined activity and is usually high on the agenda of many staff discussions. It has also created a great deal of anxiety and apprehension amongst teachers; one of the aims of this Unit is to allay some of these anxieties.

What is different between what used to happen and what we are now required to do? Is there a difference?

Some of the first documents to set out clearly the processes and strategies of assessment as it relates to the National Curriculum were those produced by the Task Group on Assessment and Testing (TGAT). The introduction to its first report contains ideas which continue to be relevant today:

> *The assessment process itself should not determine what is to be taught and learned. It should be the servant, not the master, of the curriculum. Yet it should not simply be a bolt-on addition at the end. Rather, it should be an integral part of the educational process, continually providing both 'feedback' and 'feedforward'. It therefore needs to be incorporated systematically into teaching strategies and practices at all levels.*
>
> (DES 1988)

In other words, assessment should drive the learning process – not lead it by the nose or push it in the wrong direction – and to help it do so, it should have the same criteria for achievement as the activity itself. To

It should not simply be a bolt-on addition at the end

incorporate meaningful, useful assessment into the structure or system of any educational process requires thought and planning. But first, you need to decide *why* you want to assess.

Why Assess?

The document 'From Policy to Practice', produced by the DES, contains useful definitions of several purposes of a national assessment system. It says that it can be:

> ***formative***, *in providing information which teachers can use in deciding how a pupil's learning should be taken forward, and in giving the pupils themselves clear and understandable targets and feedback about their achievements. It will also provide teachers and others with the means of identifying the need for further, diagnostic assessments for particular pupils where appropriate to help their educational development;*
>
> ***summative***, *in providing overall evidence of the achievements of a pupil and of what he or she knows, understands and can do;*
>
> ***evaluative***, *in that comparative aggregated information about pupils' achievements can be used as an indicator of where there needs to be further effort, resources, changes in the curriculum, etc;*
>
> ***informative***, *in helping communication with parents about how their child is doing; and with governing bodies, LEAs and the wider community about the achievements of a school;*
>
> *helpful for* ***professional development***, *in that the process of carrying out systematic assessment, recording attainment, and moderating the outcomes in discussion with other teachers will provide a valuable basis for teachers to evaluate their own work and to gain access to new thinking.* (DES 1989a)

We are all familiar with most of these processes and they are part of our normal, necessary classroom practice. The National Curriculum requires us to be more rigorous in the procedure of assessing and more formal in the ways in which we record these 'value judgements', yet the purposes of teacher assessment, in general, mirror well-established primary practice.

What to Assess in Science

Bouncing different balls to see what height they reach is a popular classroom activity. As a science investigation it can require children to predict or hypothesize, to understand the need for a 'fair test', to control the number of variables, to choose and use an appropriate way of measuring what happens and to record their findings. Interpreting their observations, drawing conclusions from their results and communicating these to others encourage an understanding of the properties of the individual balls and of the materials of which they are made or were bounced upon.

This example illustrates the two facets of science:

* the skills and processes that are used when carrying out an experiment or investigation
* the vast body of knowledge, both of facts and of concepts, that continues to accrue after centuries of such experimentation

Young children are obviously at the very beginning of their understanding of this knowledge and the associated concepts, although, because they have been trying to make sense of the world around them from an early age, they have developed their own interpretations, which may need challenging (see Unit 7). They have been investigating their surroundings in various ways, but now need to recognize the importance and significance of the skills and procedures that are involved when 'finding out'.

In order for you, as the children's teacher, to be able to decide just what they know and understand, it is necessary to have a picture of the progressive development of knowledge and understanding, and of the use of various skills in a scientific way. You can then match what you have found out about a child to a particular part of this continuum. The National Curriculum is probably the first document to be produced for primary teachers that describes a possible developmental framework for science.

What Else Do We Need?

Evidence The only way that we can know anything of children's thought processes, their level of understanding or their ability to use particular skills is from the evidence that they can provide for us. The nature of this evidence varies according to the activity that the child is engaged in and what it is that you want to find out about. For instance, looking at a mobile that a child has made may tell you about his or her understanding of balance and of the materials that were used. Children's drawings or written descriptions of a woodlouse can give you some indication of the development of their ability to observe; whilst talking with them about the ways in which they can find out whether these animals prefer dark or light places can tell you about the understanding they have of the significance of various experimental procedures.

So, as teachers, we need to collect appropriate evidence to help us know more about a child's thinking. This may take the form of some kind of artefact or model, or it may be a drawing or description, a recording or a game.

Observation skills It is not just 'having eyes in the back of your head' but being able to observe with a purpose and realizing the significance of what you observe. This part of the teacher's role is often underrated, and yet many of our decisions about the child's next task are based on what we have observed him or her doing up to now.

Often the most appropriate way to gain the evidence that will tell you about children's understanding of the notion of a 'fair test' is to watch them as they carry out an investigation. How do they proceed, for example, if they are testing the relative strengths of various materials? Is it fair? Are they using the same amount of each material? Is the device for breaking the material always used in the same way?

Making notes as you watch children involved in science activities can be a very fruitful way of gaining the evidence that you require.

Criteria One of the ways often employed to decide whether a cake is cooked or whether it requires more time in the oven is to stick in a metal skewer and see whether it comes out cleanly. This is not something that we can do to children, but we can use other criteria as benchmarks for their development.

This is where the National Curriculum documents are helpful. The criteria that are expressed by the Statements of Attainment and that are divided into sets, describing children at different 'levels', can be referred to when making our assessments. We can also use these criteria when planning activities, so that the intellectual challenge of an activity reflects the opportunities for assessment.

For instance, the task of making the 'best' spinner from a given selection of different kinds of paper can be designed so that the child has to 'ask questions, suggest ideas and make predictions . . . and carry out a fair test'. This will involve planning an investigation where the plan indicates that the dependent variables have been identified (time of spin/number of turns) and others have been kept the same – controlled – (size of spinner, height dropped). The ability to meet these criteria is one way of identifying, in the terms of AT1, a 'level 4' child.

Once you are familiar with the various criteria as expressed in the National Curriculum documents, you will be able to identify others for yourself that indicate 'level 2½', for instance. It is also worth remembering that children's development and learning do not follow a straight line and that your assessment of a child today may be quite different from your opinion of last week.

Planning If assessment is to be 'an integral part of the educational process' (DES 1988) and if we are continuously to assess the children in our classroom in order to decide how their learning should be taken forward, then we need to identify carefully the opportunities for assessment in our planning.

The Programmes of Study offer a framework for planning classroom work, but it is the individual Statements of Attainment (SOAs) within each Attainment Target that provide the criteria for assessment and assist you to select the opportunities for assessment.

There are fewer SOAs in the revised Order for Science (DES 1991) – 176 instead of 409 – and many of the new statements describe broader areas of knowledge and understanding than those published in the previous Order (DES 1989b). This means that each SOA has to be interpreted and 'unpicked' to provide perhaps several statements that can then be used as criteria for assessment.

In the aptly named book, *Moving Targets* (ASE 1992), the Association for Science Education suggests that teachers should not throw away the old Orders but use them to refer to when making decisions about attainment. This is a helpful publication, for not only does it identify the differences between the 1989 and 1991 Orders for Science but it also provides an analysis of the 1991 Programmes of Study, identifying the progression of each strand through the various Key Stages.

Practical considerations Many of the concerns that are voiced by teachers about assessment in their classrooms are to do with problems of management: organizing their time and that of the children, organizing the classroom, using help from other adults (parents, classroom assistants, headteachers) and other similar issues.

One way to tackle these problems is to discuss with colleagues how best to create opportunities for observation and assessment within your classrooms. You might find it helpful to refer to Chapter 8 in the book *Teacher Assessment – Making it Work for the Primary School* (ASE 1990), where it is suggested that the context provided by the classroom and the planned activities should ensure that the children

* are autonomous
* work collaboratively
* are offered open-ended tasks
* often discuss
* reflect on their learning in feedback sessions

References

Association for Science Education (ASE) (1990) *Teacher Assessment – Making it Work for the Primary School*, Hatfield, Herts: ASE.
—— (1992) *Moving Targets*, Hatfield, Herts: ASE.
Department of Education and Science (DES) (1988) *National Curriculum: Task Group on Assessment and Testing*, London: HMSO.
—— (1989a) *National Curriculum – from Policy to Practice*, London: HMSO.
—— (1989b) *Science in the National Curriculum*, London: HMSO.
—— (1991) *Science in the National Curriculum*, London: HMSO.

STIMULUS ACTIVITIES

Activity 1

Penulums

We hung a Container on. a metre stick,
We put a 200 gram Weight in the container
We Swung it. It took one second to Swing.
We tryed it With 100 grams, it still took 1 second,
We tryed it With 50 grams. It took a spit Second
We think that it be cause It is Lighter than the others.
we Made the string shorter and put 200 grams in it,
We Swung, it took a spit Second to Swing
We tryed it With 100 gram. it took 1/4 of a second.
We tryed it With 50 grams it took 1/4 of a
Secand to Swing. It was a fair test because
We swung it from the Same place every time
And We put a Weight on the String so
it did'nt move. We think that it's
the length of the String that decides
on how long it takes for a Swing
because When the String Was shorter each
Swing Was quiker.

Figure 5.1

This piece of work was produced by Anna, who is 9 years old. Read it carefully; as you do so, you may like to make notes.

What does it tell you about Anna's ability to observe and measure?

Does she really understand about 'fair tests'?

Is the last sentence an accurate interpretation of her results? How does it relate to the hypothesis that appears part way through her description: 'We think that it [is] because it is lighter than the others'?

This piece of work can provide you with a partial understanding of Anna's ability to plan and carry out an investigation. If she were sitting beside you now, what kind of discussion would you have?

How could you use this assessment to plan her next work in science?

Activity 2

Melanie's class has been investigating the way that kites fly. She made a kite to her own design and wrote about it (see Figure 5.2).

If you were assessing her in terms of AT1 Scientific Investigations, what would you say?

Does Melanie's work tell you anything about her understanding of materials, flight or balance?

What would you need to encourage her to do in order to find out whether she can plan an investigation – controlling variables and constructing a fair test?

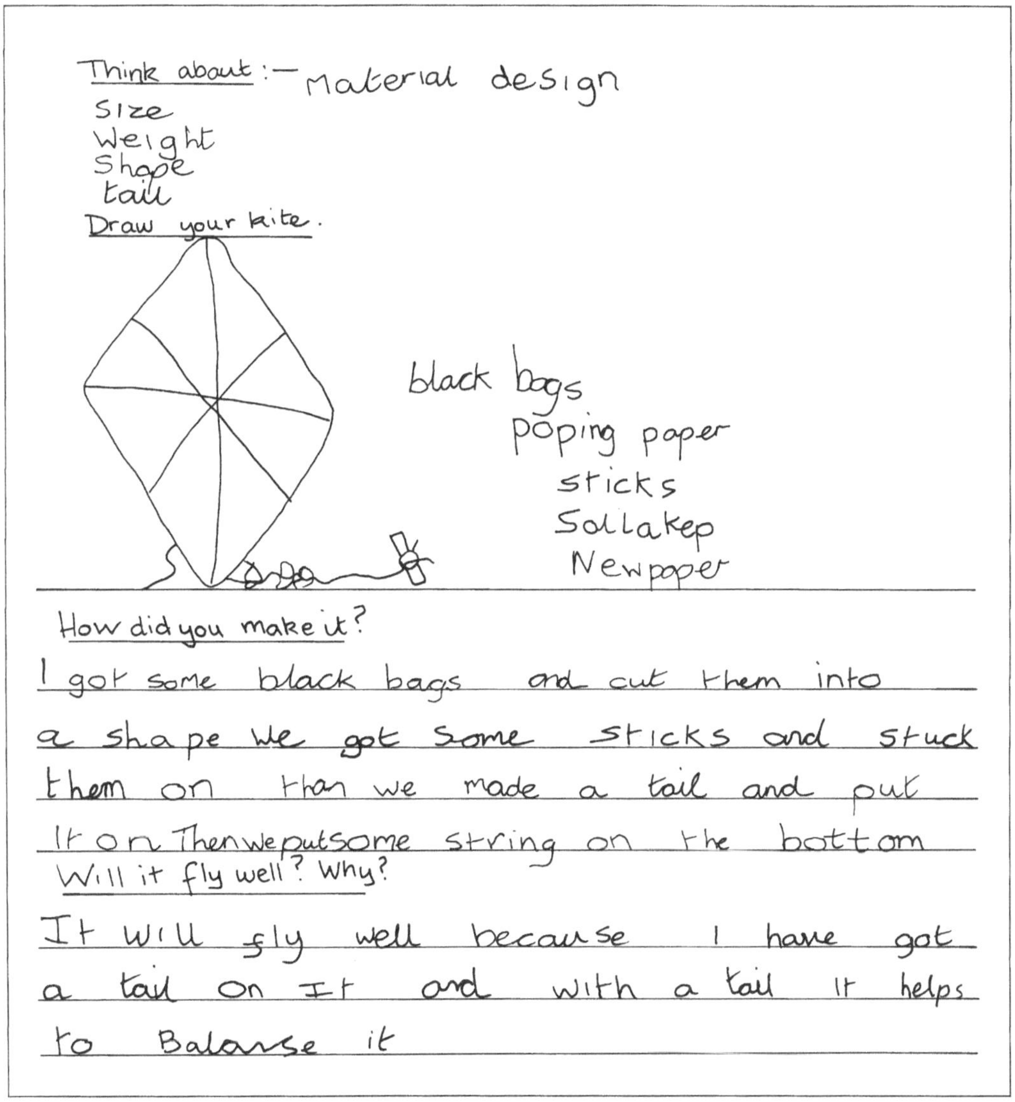
Think about :— Material design
Size
Weight
Shape
tail
Draw your kite.
black bags
poping paper
sticks
Sollakep
Newpaper
How did you make it?
I got some black bags and cut them into a shape We got some sticks and stuck them on than we made a tail and put It on Thenweputsome string on the bottom
Will it fly well? Why?
It will fly well because I have got a tail on It and with a tail It helps to Balanse it

Figure 5.2

Activity 3

You often have to create the opportunities that will enable you to assess an individual or a group of children. This Activity is designed to help you identify some ways of doing this, as well as alerting you to possible decisions that you will need to make.

Margaret Smith is planning work for her class of 7–8-year-old children. The topic for the term is 'Let's move', and Figure 5.3 shows the first stage in her thinking about the science that could be included.

She now needs to identify particular activities, bearing in mind *what* she wants the children to learn and understand, and *how* she is going to assess their progress. In your groups, discuss the kinds of activities you think Margaret should choose, and decide which SOAs it would be appropriate for her to assess through this work – don't forget AT1.

What will the children have to do in order to demonstrate that they have achieved these SOAs?

Will Margaret be looking at the ways in which they carry out the task, what they produce at the end, or will she need to talk to the children after they have completed the activity?

How many children do you think that she could assess in one session?

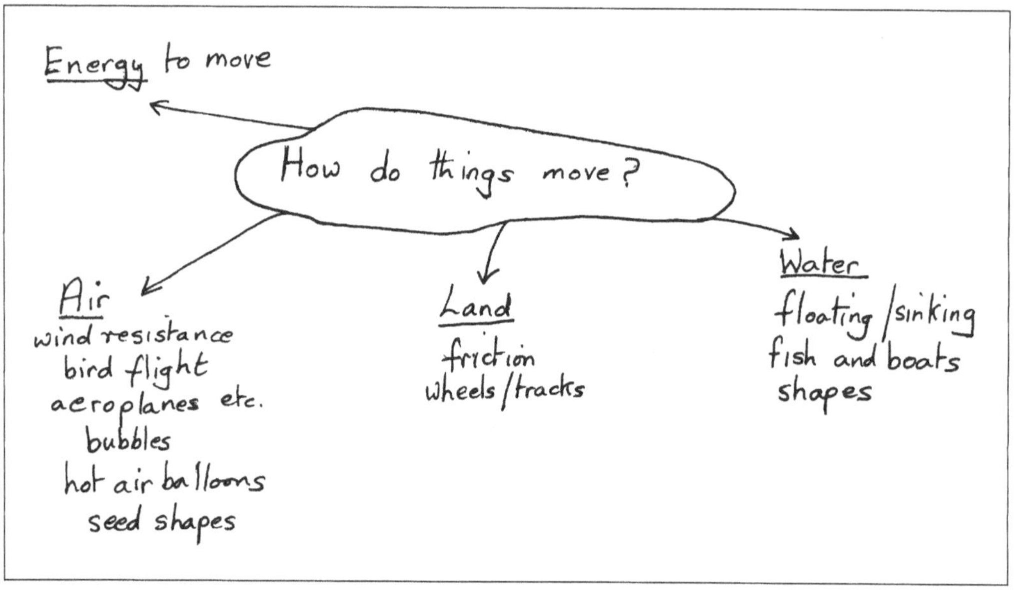

Figure 5.3

Activity 4

Working in pairs, choose one activity from the following list. Then, choose one or more Statements of Attainment from AT1 that you think would be addressed by this activity.

1. How good are we at recognizing tastes – with our eyes covered? holding our noses as well?
2. Woodlice are usually found under stones or logs. Why do you think this is? How could we find out?
3. We need to send a parcel to a school in Ethiopia. We have a selection of different papers which could be used to wrap it up. How can we find out which is the best paper for this purpose?
4. Grapefruit are usually sour and, when eaten raw, are often covered in sugar. Does the sugar 'take away' the sourness or does it merely 'mask' the taste?

Decide how you would explain the task to a group of children in your class in such a way that when they carried out the activity they would demonstrate whether or not they had attained these SOAs.

What are the implications of this kind of work for the organization of your classroom? What will those children be doing who are not being assessed?

You may like to discuss your ideas with other teachers in your group, before repeating the exercise with another activity from the list.

CLASSROOM ACTIVITIES

Activity 1

Choose a selection of children's recording in science from your own classroom – they may be drawings or pieces of written work.

Focusing on one piece of work at a time, decide which SOAs and ATs it is related to, and note where you can confidently assess the child's level:

a. from a drawing or written work alone
b. from your observations and knowledge of the child, as well as the written work.

Is there anything that you have been unable to assess in these ways and, if so, what future plans for assessment will you make?

Activity 2

Tape-record two or more children in your classroom whilst they are engaged in an investigative activity, and listen to your recording carefully.

Decide which SOAs and ATs the activity is related to, and note where you can confidently assess each child's level:

a. using only this recording
b. using your observations and knowledge of the child as well.

Is there anything that you have been unable to assess in these ways and, if so, what future plans for assessment will you make?

Repeat Activities 1 and 2 as often as you can with other examples, and include those that have different end points – for example, a model, a drawing, a game or a display.

You may like to discuss your findings with a colleague. It is especially helpful if you can 'team-teach' a class, involving someone else in the planning of the activities and in assessing the children.

Activity 3

Analyse the plans that you are currently employing with your class this half-term in the following way:

a. Which activities are going to provide you with the opportunities for assessment? Identify the SOAs
b. What will the children have to do in order to achieve the SOAs that you have identified?
c. Will you be assessing the ways in which they carry out the task and what they produce at the end, or will you need to talk to the children after they have completed the activity?
d. How many children do you think that you will be able to assess in one session?
e. What are the implications of this for your classroom organization? How are you going to manage the children who are not being assessed?

Activity 4

Choose a science activity that you plan to carry out in your classroom next week, and identify a SOA from AT1 that is appropriate for this activity.

Explain the task to the children in such a way that, when they carry it out, you are able to see whether they have attained this SOA. You can repeat this with other activities, using SOAs from different ATs.

REVIEW SESSION

1. Choose a partner who has placed at least two children at the same level in AT1 as two from your class. Describe and discuss with one another the work of these children.

What is it that characterizes them as, say, 'level 3'?

Do your criteria for 'level threeness' match, or were there inconsistencies?

Are you both interpreting the relevant SOAs in the same way?

Join up with another pair of teachers who are discussing children at the same level as yours, and ask these questions again.

Now, talk with teachers who have been discussing a higher or lower level than you, and decide what it is that differentiates one group of children from another.

2. Share with your group the analysis that you made of the assessment of children's work and of your planning (Activities 1 and 2, 3 and 4 in 'Classroom Activities').

What did you each find that you needed to do in order to achieve a more complete understanding of the levels of attainment of the children in your classes?

Were there things that you did not manage to achieve? Do you know the reasons for this? For example, 'I didn't assess enough children, because it took longer than I expected to assess each child.'

3. Discuss within your group the problems that you have found with assessment, teaching and planning. For example, were you able to fit into your planning enough activities to address a wide range of SOAs from AT1? Did you find that there was a tension between 'teaching' and 'assessing'? Were you able to teach a child at the same time as assessing him or her?

Can you suggest some strategies that you need to develop in order to overcome those difficulties that you have identified?

UNIT 6

INTRODUCING A MULTICULTURAL DIMENSION INTO SCIENCE

LEADER'S GUIDELINES

This Unit should help teachers to be more aware of the need to introduce a multicultural element into their science work whatever school or age they are involved with, and to gain confidence in tackling their planned activities from a multicultural perspective.

Briefing Paper

This should be distributed after the first four Stimulus Activities have been completed, to be read before participants undertake Stimulus Activities 5–7. The paper makes reference to a number of background texts which are an important source of ideas; it would be helpful if you made some of these available to participants during the Unit.

Stimulus Activities

There are seven activities altogether:

1–4. These are intended as short orientation activities, and could all be completed in one introductory session of $1\frac{1}{2}$ to 2 hours. The Briefing Paper should be distributed at the end of this session.

5–7. These are planning activities, intended as precursors to classroom work. They can all be undertaken within one good session of about 2 hours, if the intention is to achieve outline planning rather than in-depth plans. Alternatively, participants can be briefed on these activities prior to the session, and be asked to come with ideas and skeleton plans for working on with other participants.

1. The purpose of this activity is to help participants see the multicultural potential of the topics they are teaching. It would be helpful if you as leader had studied the Planning Checklist in the Briefing Paper before this activity, to help them focus their thinking.
2. The focus of this short activity is on the values dimension of science. If participants have difficulty thinking of ideas, you may ask them to consider the relevance of our science National Curriculum to the Ethiopian village community mentioned in the example in the Briefing Paper. What science learning would be of most value to these children?
3. The questions in this activity are deliberately chosen to help participants articulate their own values. There is clearly a political and economic dimension – the point being that these are inescapable in science. You may wish to write each question on a separate card,

and ask a pair of participants to consider one question before reporting their thoughts to the group.

4. Lack of resources is often cited as a reason for not tackling multicultural education. The point of this activity is to show teachers that they do have access to resources, often in unexpected places, and that these don't have to be 'exotic artefacts' but can be simple yet valuable things like food labels. If it seems appropriate, you might ask the group to bring a collection or a valuable resource to the next session, to create a group display.
5. Planning a second language activity. Prior to this activity you should find out which participants have some skills and confidence in a language other than English, and group participants around these. Complete fluency is not required. Groups of three or four would be ideal. The draft worksheets produced in the activity of a science task in a language other than English should be circulated briefly at the end of the session, for other groups to evaluate. You may wish to copy and distribute good examples as part of your resource bank.
6. Planning a group activity. This should be undertaken in small groups. Spend 15 minutes or so on outline planning of the activity, then ask groups to present and compare their plans. Refer participants to the boxes on Communication Skills and Linguistic Demands in the Planning Checklist, and to the points about the value of collaborative group work in assisting language development.
7. This activity is best undertaken as a whole group session, to maximize the ideas generated. You could, if you wish, ask sub-groups to look at one factor each before opening up the whole-group discussion. As leader, you might generate a group plan on a flip chart, as you go along. At the end of this activity, you should help the group to brainstorm a list under this heading: 'The things we still need help with are. . . .' You might then invite each participant to agree to do something about at least one of these, before undertaking the Classroom Activities.

Classroom Activities

There are three Classroom Activities:

1. Involving children in accumulating multicultural resources.
2. Introducing a science session in another language.
3. Planning and carrying out a science activity with a multicultural content.

1 and **2.** These can be undertaken after the first session of Stimulus Activities has been completed. Activity 1 may be an on-going activity which takes time to complete, so you should allow two or three weeks at least before holding a review session. Activity 2 is short and can be undertaken at any time; it may be best to have a Review Activity on this soon after it is completed, perhaps at the beginning of the second session of Stimulus Activities.

3. This activity may also take time to plan a d to find a suitable slot for it in the scheme of work, so again it may be best to review this after several weeks, alongside Activity 1. Participants may need help in arriving at a suitable choice of focus for this activity; if so, you might refer them to the examples provided in Peacock (1991: ch. 5).

Review Sessions

You will probably need one longish session after several weeks to review Classroom Activities 1 and 3, and one short session linked to some other meeting of participants to review Classroom Activity 2.

In reviewing Classroom Activity 1, focus on the availability of resources and sharing information about access. You might encourage participants to find ways of establishing a shared multicultural resource bank in relation to topics of common interest.

The focus in reviewing Classroom Activity 3 is on the multicultural content of the activity. You should use the Planning Checklist in the Briefing Paper to provide criteria by which to judge the appropriateness of the activities, and encourage participants to evaluate by cross-questioning one another. They should all bring their plans and evidence from the activity to the Review Session.

After Classroom Activity 2, invite participants to talk about their own and their children's feelings as a result of this experience. It is important to encourage teachers to put their conclusions into the context of bilingual children in Britain and in the rest of the world, and to think about how practical group work in science can be used as an important vehicle for language development, as this should inform their subsequent planning and classroom activities.

BRIEFING PAPER

Principles

> *Given then that so many of the principles by which Science is most effectively learned in primary schools are also those of anti-racist education, surely primary Science has no excuse for not being in the vanguard of multicultural/anti-racist practice?*
>
> (Troyna and Farrow 1991)

No excuse, maybe; but in practice, research has suggested that primary teachers and science teachers have been shown to be the two groups within the teaching profession least likely to adopt multicultural approaches to their work with children. So what are these common principles, and how can teachers be helped to apply them in their teaching?

Some of the principles that have been suggested as fundamental to anti-racist education are as follows:

1. the adoption of a non-didactic approach to learning;
2. value being placed on cooperation and collaboration in learning;
3. the acceptance of a range of differing and possibly conflicting points of view;
4. the acceptance of rational and independent argument.

And it is true that these are mirrored in effective science learning, when we note that good primary science involves the following:

1. not simply inculcating a body of 'right answers' but requiring children to test ideas against the evidence of their senses, collected by observation and measurement;
2. requiring children to undertake practical investigation in groups, in which they jointly plan, carry out and discuss their work, and present an agreed conclusion;
3. in attempting to explain conclusions, children will come up with different, conflicting hypotheses, which are all equally valid until tested;
4. testing hypotheses by collecting evidence means accepting the independently gathered evidence of others, not simply the evidence that suits our predictions or prejudices.

So practical, investigative science is an ideal vehicle for establishing a multicultural and anti-racist dimension into your work. But where it is not yet happening, how do you get started?

Starting Points

We can begin by thinking of science as an activity:

* carried out by people
* which has relevance to all aspects of human life
* which involves processes of language, observation, thought and discovery
* which children enjoy doing

This suggests several things a teacher can do to help children, namely:

* challenge the stereotype of scientists (as white, western men in white coats) by showing that we all do science
* acknowledge and introduce the cultural diversity of our local, national and global community into our science work
* value the similarities between people of different backgrounds, as well as valuing the differences in language and culture

In other words, it is not just a matter of *what* we teach (investigation skills, concepts) or *how* we teach it (practical work in groups), but also of the *context* in which we plan science work. This means children's own experiences in different social contexts – diet, nutrition, energy, clothing, health, music, conservation – all need to be tackled in ways which accept the richness and diversity of our cultures, rather than by taking a narrow view of 'correctness' in these matters.

Two quick anecdotes might help illustrate this point.

A teacher once told me that he had carried out a survey with his class on 'What we ate for tea'. Being a multi-ethnic classroom, many children didn't have a clear concept of 'tea' or 'teatime', but they knew what 'English' kids had for tea, so they all wrote down 'fish and chips' or 'burgers', even though some were vegetarian.

An Indian girl who showed very little interest in her school-work and hardly talked in class, suddenly became animated in a science lesson on fuels when the teacher brought in some dried cowdung to burn. 'It smells just like home, miss!' she said with a smile.

This point about context is not only valid in multi-ethnic schools, either. The Association for Science Education, in its policy document on *Race, Equality and Science Teaching* (1990), makes the point in this way:

> *We believe that a science curriculum which ignores the above issues will disadvantage all children: white children because of the impediment caused by the unchallenged assumptions and attitudes they hold about black and other minority communities; and children from black and other minority groups because their progress will be impeded by inequalities in their access to science provision. Furthermore, all children will be denied the enrichment brought to science by the multicultural dimension.* (ASE 1990: 1)

It is therefore worth while thinking about this context dimension in choosing topics and materials. We can start by looking at ourselves, and can move on to include the homes children live in; the clothes they wear; the food they eat; the festivals they celebrate; the toys and games they play with; the tools, utensils, machines and instruments they use; the way they travel; the way they keep warm or cool; and the work they and their families do.

If this sounds daunting, and you feel you don't know enough about the clothing, homes or festivals of other cultures, try learning alongside your class. A simple way to start this off is to go to a large supermarket and buy a range of as many different fruits as you can find, especially those that are unfamiliar to you. You could, if you like, do the same with breads or vegetables, or anything where an unusual variety is available. Bring these into class and investigate them along with your children. Get them to raise questions about the collection – there will be no shortage of these – and let groups investigate questions which interest them. Admit you don't know all the answers; and carry out some of the tests with your children. Where fruits are concerned, for example, you can predict what they will look like inside; how many seeds they have; the thickness of the skin; how you can tell if they are ripe; and you will probably find that some children are pleased to be able to supply you with answers to some of your questions. They will be motivated by your own enthusiasm to find out, and their self-esteem will be boosted by using their knowledge and helping their teacher. And you will all be learning science! A detailed illustration of how this worked out in one teacher's classroom can be found in Peacock (1991: ch. 2, 'The multicultural dimension in the science of food', by Chris Hannon).

But please don't think that none of this applies to you if you don't teach in a multicultural school. Pursuing objectivity through science is a good habit for all children to develop. The alternative may simply be the re-affirmation of prejudices which, when not countered by evidence, start to sound like truth, so that children only hear and see the evidence that they want to see, and that confirms their prejudices.

Take as an example the grouping of children by skin colour or eye colour or hair colour or any other physical attribute, often carried out under a topic like 'Ourselves'. So often we produce bar charts or pie charts which use clear categories like 'brown eyes . . . blue eyes' which creates an illusion that people fall into clear types (which, if we are not careful, can then be related to other, sometimes derogatory, characterizations). But as soon as you try to arrange a class of children objectively, scientifically, along any single dimension such as eye colour, you observe that in fact there is continuous variation in colour, and that there is no obvious, definitive place to draw the lines that allow people to be put into categories. So our 'types' are not really scientific.

Science and Language Development

The specialized use of vocabulary in science can be a barrier to understanding: ask your class to explain the meaning of words such as 'energy', 'work', 'force', 'hot' or 'animal', and you will notice that there is a wide range of everyday and scientific meanings for each of these. At the same time, science is an important vehicle for language development, because it involves communication (instructions, questions, information gathered, observations) about tangible, concrete objects which have to be manipulated and described, in order to predict and explain and name things. So talk and discussion are essential starting points in children's development of science concepts, especially for bilingual learners. And it is therefore important that the diversity of children's linguistic as well as scientific experience is properly valued and built on.

This means using a wide range of learning strategies; allowing children to explore and explain their own ideas, initially in their own language if necessary; and intervening to introduce the scientific use of terminology when you can see that a child has acquired the concept, but hasn't yet got the name. It also means accepting children's misconceptions, and using them as starting points for science learning.

For example, a student teacher I supervised recently was letting children push a football down into a tank of water, during a lesson on buoyancy, and asking them 'Can you feel the water pushing up?'. 'But miss', said one Bangladeshi girl, 'water hasn't got any life to push with.' This was the starting point for a fascinating discussion about the scientific idea of 'push' (force) and about the contexts in which you can use the word 'push': can only living things push?

Science utilizes language for a whole range of functions: identifying, questioning, locating (in space and time), quantifying, describing, interpreting, explaining, reporting, reasoning, sequencing, estimating, projecting, instructing. Being multicultural about science means, therefore, removing the barriers provided by inaccessible language. The difficulties and differences between everyday speech of children and scientific use of language must therefore be discussed with children as they arise, and wherever they arise, in schemes, books and worksheets, as well as in your own language use. The Planning Checklist at the end of this paper summarizes the main points that you might need to bear in mind when planning multicultural science work for children. If you are not fully aware of what all this means for a bilingual child, Stimulus Activity 5 and Classroom Activity 2 are intended to help put you in their shoes.

Criteria

The remainder of this paper provides two checklists against which you can compare your efforts to make your science teaching more multicultural. The first is a list of ideas by which you can help children become multicultural/anti-racist as they become scientific.

1. Encourage all children to work together in groups, and not simply with friends.
2. Start from children's knowledge, interests and understanding of any particular topic.

3. Place science in a social, moral, political and economic context – for example, by
    a. the children bringing in newspaper or magazine articles on wider issues (e.g. *Education Guardian*)
    b. the children watching and discussing TV programmes ('Living Planet', 'Horizon', etc.)
    c. the children discussing news items with a science dimension (e.g. hurricanes).
4. Use artefacts, plants, foods, tools, etc. from different cultures in scientific investigations.
5. Encourage children to bring in familiar objects.
6. Value other people's customs and encourage children to do the same.
7. Let festivals and celebrations be a starting point for scientific activities.
8. Ask children to look for similarities with one another, not just differences.
9. Teach and apply the notion of Continuous Variation. *Note*: variation in skin colour is an example of Continuous Variation, as all skin colours can be seen to grade one into another. There are no clear-cut boundaries between 'black', 'white', 'yellow', etc.
10. Encourage children to talk about their science activities and ideas.
11. Allow young children to use their mother tongue when discussing their scientific ideas.
12. Value non-verbal communication (drawings, diagrams, dramatization, display) of scientific ideas and activities.
13. Challenge racism yourself, and teach children to do so.
14. Avoid notions of 'correctness' in taste, dress, habits, etc.
15. Let children tackle 'real' science problems at their level.
16. Use scientific ideas to help children understand customs in other lands (e.g. house design, diet, dress).
17. Involve the community in formulating multicultural ideas and policy in science work (e.g. by discussing these issues with parents and governors).
18. Make sure children of all backgrounds are encouraged to present their scientific findings to others in the school.

(Adapted from Wyvill 1991)

The second checklist is intended to help you with the planning of programmes of science work with a multicultural dimension, although it is applicable to any and all topics, not simply science. A planning sequence is suggested (boxes 1–10, in the centre of the diagram) and checklists of things to think about at each stage are indicated in the larger boxes to left and right of the sequence, covering science, language and multicultural concerns.

PLANNING CHECKLIST FOR MULTICULTURAL SCIENCE

1 What TOPIC from your scheme of work have you chosen?

↓

2 How many LESSONS/WEEKS do you expect it to last?

↓

3 Which INVESTIGATIVE SKILLS (AT1) will children practise?

AT1

- Raising scientific questions
- Exploring everyday experiences
- Working carefully and safely
- Investigating and reporting systematically
- Using equipment
- Measuring
- Progressing from qualitative to quantitative data
- Identifying and controlling variables ('fair tests')
- Searching for patterns in data
- Interpreting data

↓

4 Which KNOWLEDGE and UNDERSTANDING (ATs 2–17) will they cover?

AT2 Life and living processes
AT3 Materials and their properties
AT4 Physical processes

↓

5 What ACTIVITIES will children do? (in OUTLINE ONLY at this stage)

↓

6 What COMMUNICATION SKILLS will these activities develop?

- Expressing needs, understanding instructions
- Naming, labelling, describing
- Comparing, classifying, attributing
- Predicting, anticipating, estimating time
- Locating objects in space
- Estimating distance, direction, size
- Identifying question words
- Sequencing, ordering logically
- Recognizing relationships, analysing evidence
- Suggesting reasoned explanations
- Projecting into unknown situations
- Creating from experience

↓

7 What LINGUISTIC DEMANDS will they make (especially on E2L children)?

- Using specialist vocabulary, symbols, abbreviations, conventions
- Giving and responding to instructions
- Asking and answering questions in groups
- Speaking freely and audibly
- Making a well-ordered account of an event or process (oral/written)
- Using appropriate medium of communication (English/other)
- Using sources of information

↓

8 What MULTICULTURAL CONTENT is there in the activities?

- Are topics relevant to all cultures?
- Are there specific activities which might motivate ethnic-minority pupils?
- Is interdependence stressed?
- Are similarities examined as well as differences?
- Are processes (e.g. food preservation) related to cultural context?
- Is the contribution of other cultures stressed?
- Are global issues such as health or conservation dealt with adequately?

↓

9 How can IMPLIED RACISM be avoided in the approach taken?

- Are community attitudes taken account of (e.g. to food, sex, dress)?
- Do the book/materials used reflect cultural diversity?
- Do they promote linguistic diversity?
- Is stereotyping avoided?
- Are the origins of science in other cultures acknowledged?
- Are the practices of other cultures presented as backward/deprived?

↓

10 How do you intend to START?

References

Association for Science Education (ASE) (1990) *Race, Equality and Science Teaching*, Hatfield, Herts: ASE.

Barnfield, M., Comber, M., Dyble, L., Farmer, M., Hagues, S., Hughes, S., Martin, T., McFarlane, C. and Moore, A. (1991) *Why on Earth? An Approach to Science with a Global Dimension at Key Stage 2*, Birmingham: Development Education Centre.

BBC Enterprises (1988) *Instruments of the World*, London: BBC Enterprises. Wide range of suggestions for making/improvising/playing musical instruments, from a science and technology approach.

Button, J. (1989) *The Primary School in a Changing World*, London: CWDE.

Centre for World Development Education (CWDE), Regent's College, Inner Circle, Regent's Park, London NW1 4NS; tel. 071-487 7410. A wide range of relevant materials, including free catalogue.

—— (1990) *The Water Game* (computer simulation), London: CWDE.

New Internationalist (1989) *NI Peters Atlas*, NI Publications, Oxford OX1 2BR.

Oxfam (1989) *The World in a Supermarket Bag*, Oxford: Oxfam (also from CWDE).

Peacock, A. (ed.) (1991) *Science in Primary Schools: the Multicultural Dimension*, London: Macmillan Education (now from Routledge Education).

Troyna, B. and Farrow, S. (1991) 'Science for all? Antiracism, science and the primary school', in A. Peacock, *Science in Primary Schools: the Multicultural Dimension*, London: Macmillan Education (now from Routledge Education).

Wyvill, Barbara (1991) 'Classroom ideas for antiracism through science in primary education', in A. Peacock, *Science in Primary Schools: the Multicultural Dimension*, London: Macmillan Education (now from Routledge Education).

STIMULUS ACTIVITIES

Activity 1

Make a list of all topics being covered by participants in the current school year.

Activity 2

Here is a list of just a few topics suitable for introducing a multicultural dimension into science work with primary school children:

Fruits and seeds
Cooking
Food preservation
Tools and utensils
Musical instruments
Energy and work
Houses
Clothing
Soil
Wind and water

Activity 3

Brainstorm a list of reasons why the topics above are appropriate to a multicultural approach to science.

Activity 4

Now look at the list you made in Activity 1. What areas of overlap are there between the two lists?

Activity 5

'Language and number systems are different in different cultures, but science is the same wherever you are.'

Brainstorm two lists, one of reasons why you might agree with this statement and one of reasons why you would disagree.

Activity 6

Think about, then discuss, at your own level, your answers to some of the following questions related to the topic 'Food':

Why are many people short of food?
Which foods do people value most?
Who owns the land on which food is grown?
Why do many people live on poor land?
Who has access to better land, and why?
How important is climate in growing food?
Why is it important to grow cash crops?
How is the price of a cash crop (e.g. tea) decided?
Who grows the food?
How much of our food do we grow?
Who is dependent on whom for what?

(Adapted from Barnfield *et al.* 1991 – see page 68)

Make a list of things you need to know in order to answer these questions satisfactorily.

Activity 7

Lack of suitable resources is often given as a reason for not including a multicultural dimension in science.

Go through the following headings and add to each one some more examples of resources you could use in a multicultural way.

	Example
The children	those with 'roots' in other places
Parents, visitors	who have travelled/worked in other cultures
Artefacts	musical instruments, cooking utensils
Collections	food labels, photographs, fruits
Museums, centres	Development Education Centres, CWDE*; churches, mosques, temples, gudwaras, synagogues
Books	BBC Enterprises, *Instruments of the World**; stories
Video and TV	BBC/ITV for schools
Maps, atlases, globes	The Peters Projection* maps
Newspapers	*Education Guardian*
Games	Oxfam, *The World in a Supermarket Bag**
Software	CWDE, *The Water Game**
Free leaflets	Oxfam, Friends of the Earth, ITDG
Links	pen pals, twinning, UKOWLA

* Sources of these materials and organizations are given in the references at the end of the Briefing Paper (p. 68).

Activity 8

Choose a simple science activity for which you can produce a one-page worksheet, and make a worksheet in a language other than English, using the expertise of those members of the group who are reasonably fluent in a second language. Keep the language component short and simple, and use diagrams and illustrations where possible.

Activity 9

You are doing some science work with a Year 4 class on properties of materials, and have planned one activity on 'Which ball bounces highest?'.

A child with no English joins your class at the beginning of the week prior to this activity. Plan the activity so that he or she will be able to participate effectively, understand what is going on and learn some science.

Activity 10

You have a class of Year 6 children who are all white, native speakers of English, with little experience of other cultures.

Plan an activity within the topic 'Homes' relating to the science concept on insulation (keeping warm, keeping cool). Make sure it deals with at least one of the following factors in a multicultural way:

* materials used to construct houses (roof, walls, floor)
* climate differences (hot/cold, wet/dry, windy/calm)
* appropriate technology for construction (thatch, brick, steel)
* ways of life (settled, nomadic)
* cultural factors (religion, kinship, community)

Make a list (for the Review Session) of aspects of the activity with which you think you need help because of lack of adequate information.

CLASSROOM ACTIVITIES

Activity 1

Refer back to the list of resources you generated in Stimulus Activity 2.

In one of your topics, involve your class in accumulating multicultural science resources relating to one or more of these headings. For example, they might:

make a collection which included articles from different cultures and countries;

write letters inviting visitors from other cultures, or establish a link with a school from another cultural background, relating to something of scientific interest (clothing, musical instruments, cooking, etc.);

produce a display after researching a topic from a multicultural perspective, such as 'labour-saving devices', 'clean water', etc.

Activity 2

At a suitable time (maybe the beginning of a topic) introduce a science session in another language. Spend only 5 or at most 10 minutes on this, and make sure it involves materials you can show the children, diagrams and illustrations, and other things which can provide a context. It does not matter how accurate your command of the second language is (provided no child is fluent in it!), and it does not matter which language you use; you could almost make up a language, for the purpose of the exercise. The point is for children to empathize with those (the majority in the world) who receive their schooling in a second language.

If you find this too daunting, you can achieve the same effect by asking a colleague fluent in another language to translate a simple worksheet for you, then give it to your class in the second language. For added effect, you can give one group the same worksheet in English, without telling the rest: this will heighten their awareness of the advantage that first-language learners have. You may be able to use the worksheet which you developed in Stimulus Activity 8 for this purpose.

Activity 3

Use the ideas you developed in Stimulus Activity 10 to plan and teach an activity for your class. Try to keep the science focus as uncomplicated as possible by concentrating on one skill only from AT1, and one concept area: for example, if you are doing work related to Materials (AT3) (on the theme of homes), you might focus on observation of how fabrics absorb water. Then try to make sure the activity takes account of the criteria listed in the Briefing Paper (that is, those that appear relevant from the checklist but especially those related to Stages 8, Multicultural Content, and 9, Implied Racism). When you have carried out the activity, bring your plan and some evidence used in evaluating the activity to the Review Session, and be prepared to explain the multicultural content of the activity to other participants.

REVIEW SESSION

1. Where possible, bring your resource collection from Classroom Activity 1 and share it with other participants. Discuss the effort that went into putting the collection together, and whether it was worth while. As a group, consider ways of making such resources available to one another. If you were putting together a 'Multicultural Resources Bank' on a topic of current common interest, what would it need to include, and how would you acquire these things?

2. How did you feel whilst teaching a lesson in a language children didn't understand? Were your feelings similar to those of other participants?

How did the children respond? What can you learn from their responses about how you would adapt your teaching if you had bilingual children in your class?

In what ways does practical group work help bilingual children?

3. Take it in turns to present the activity you undertook in Classroom Activity 3, and to explain how it meets the criteria in the Planning Checklist.

In evaluating other participants' activities, use specific criteria from the Planning Checklist to cross-question them about how an issue of concern to you was taken account of.

Provide constructive comment to colleagues on how their activity might be modified in order to take better account of key criteria.

UNIT 7 GETTING TO KNOW CHILDREN'S IDEAS IN SCIENCE

LEADER'S GUIDELINES

This Unit asks teachers to consider three main questions about children's science ideas:

* how can we find out what our children's ideas are?
* what common ideas and misconceptions do children have about science concepts?
* how do we respond to these ideas?

The approach underlying the Unit is that it is preferable to deal with children's misconceptions by raising questions which will engage them in more investigative activity, so that they can modify their ideas on the basis of experience. Trying to 'correct' misconceptions through explanations alone is often not effective.

Briefing Paper You should distribute this at the end of Stimulus Activity 1, so that it will help participants to reflect on their ideas emerging from the activity, and prepare them to undertake similar reflection in subsequent activities.

Stimulus Activities Activities 1, 2 and 3 should take about 1½ hours. Activity 4 will need about 45 minutes for viewing and discussion.

1 and **2.** These involve participants articulating their own ideas followed by a discussion. They should compare their analysis with the ideas in sections 1 and 2 of the Briefing Paper.

3. Analysis of children's ideas as represented in their drawings and verbal explanations. The crucial stage is the identification of the appropriate response (Alternative B in the case of the child reasoning why a plant has a certain number of seeds) and the reasons why this is the best (the child has to reject her existing hypothesis because of the evidence of her observations, requiring her to suggest a new, more appropriate theory).

4. Using the observation techniques (as described in the Briefing Paper) on a commentary-free video extract of children doing investigative science.

You should ideally use a short extract made in one of your classrooms where the context is familiar. As a substitute, you could choose a suitable extract from the 'Curriculum in Action'

videos, the ILEA Primary Science videos, or any other example which does not have an explanatory commentary. Participants should initially make their observations individually before sharing them with colleagues.

Classroom Activities

1. Examining children's ideas/misconceptions during a practical science activity and recording observations in some detail.
2. Examining teachers' own explanations of science ideas to children and recording observations in some detail.
3. This follows on from Stimulus Activity 4. Its purpose is to practise and refine observation skills which can be used during 'normal' classroom working to identify children's science ideas and, of course, to assess children simultaneously.
4. Teachers' articulation of their findings. This can be undertaken as an on-going activity throughout the period of the Unit, as a check on understanding. No individual teacher need attempt more than one of the ideas suggested, but it would be helpful if you could arrange for different teachers to undertake different ideas.

Review Session

1. Reviewing Classroom Activity 1.
2. Reviewing Classroom Activity 2.
3. Comparing findings from Classroom Activities 3 and 4.
4. Summarizing what you have learned and what you still need to find out.
5. Reporting back on reading previously undertaken.

Activities 1, 2, 3 and 4 could be completed in a single session, provided that participants have a record of their observations. The background reading for Review Activity 5 can be undertaken at any time during the Unit: participants should be alerted to the need to carry out the reading at the beginning of the Unit.

The timing of review activities depends very much on how quickly participants are able to complete their observations; it might be most useful to have a short session dealing with Activities 1, 2 and 3 as soon as most participants have done the required observation, and to leave Activities 4 and 5 until everyone has had ample time to reflect on these and digest the reading.

BRIEFING PAPER

1. How Do We Respond to Children's Ideas?

When children first come to school, they already have ideas about many things, which they articulate by talking to one another, to teachers and to parents. When we listen to these ideas, we soon realize that some of them don't coincide with our own 'adult' ideas about things. We can respond to this in many different ways. Here are the two simple examples from Stimulus Activity 1:

> ADULT *(showing photograph of Big Ben)*: What is this a picture of?
> FIRST CHILD: It's Big Ben.
> SECOND CHILD: No it's not, Big Ben is a penguin.

> ADULT *(beating a drum with a stick)*: What makes the sound?
> CHILD: As you beat the drum sound echoes like I said and then comes up and seeps through the skin on top.

How you respond to such a situation as a teacher obviously depends on your knowledge of the individual child in question. But you are also aware that the way you respond to such situations is usually decided on the spur of the moment, with little time to reflect. What this Unit tries to do is help you take account of children's common misconceptions in science, while helping them move towards more useful concepts.

There are basically two different ways in which a teacher is likely to respond to statements of the kind quoted above. One way is to *correct* the child's response, by providing your own version: 'No, not quite, it's because . . .'. The other is to *ask a further question* which will probe further into the child's own ideas, and necessitate his/her doing further investigation: 'So what does the drum skin do when you hit it?'. One reason teachers without a science background often give for their anxiety about science is that they will not be able to respond in the first way, because they don't know the answers. However, in this Unit, you will be encouraged to use the second way of responding – by asking a question which leads to further ideas and investigations – not simply to get you out of this dilemma, but because it is the best way for the children. The first approach conveys the message that science is a body of knowledge which is inside teacher's head, and the secret is how to get it out painlessly. The second approach takes the line that we only learn science by 'doing science'; by practical investigation, mediated by a teacher who understands.

2. What Science Ideas Do Children Have?

Before you can think about what ideas children have in science, you need to have a clear notion of what 'science' is. For example, look again at the two examples above, about Big Ben and the drum. Are these about science, or not? You could probably argue that the second one is clearly science, because it is about sound and how a sound-producing machine works: there is a scientific concept involved (vibration). The first example, it might be argued, is also science, because it is about penguins, which are living things: but there is no scientific concept involved, only the recognition of a photograph. The child doesn't think that the building looks like a penguin: the child simply 'knows' a penguin by the name of Big Ben, so the issue is more one of logic than science.

So the context in which an idea is expressed may be scientific, but that does not necessarily mean that the child's idea is scientific. It is generally agreed that science consists of (1) a body of knowledge, concepts and principles on the one hand, and (2) a set of processes and skills used to acquire and make sense of knowledge on the other. 'Doing science' involves testing our ideas (concepts) against objective evidence (obtained by using the processes and skills).

In trying to decide whether an idea is to do with science or not, it is worth asking yourself, in each case:

* what is the *concept* or *principle* involved?
* what is the *process* involved?

The science *concepts* relevant to teachers and children are described in ATs 2–4 of the National Curriculum, whilst science *processes* are those set out in AT1.

The way you respond to a child's idea depends very much on your own understanding of the same idea or phenomenon, so it is useful to examine your own ideas about scientific concepts. Research by the Children's Learning in Science Project (CLISP), amongst others, has shown that childhood misconceptions can survive into adulthood: even graduate physicists can retain misconceptions about science ideas! So don't be put off if you find that your own ideas are hazy or unscientific. Most people are in the same boat!

Stimulus Activity 2 asks you to examine your own scientific ideas. The questions there were chosen because they relate to areas of common misconception amongst children and adults. For example, in relation to their experience of seeing, children often believe that their eye is in some way active, 'sending out' light which then comes back from the object seen. If we are aware of these common tendencies, it can make it much easier to plan science work for children in such a way that they will eradicate their misconceptions (or 'alternative frameworks', as some authors call them) and replace them with more appropriate and useful concepts.

This activity probably won't do your ego any good! However, be reassured by the fact that most people have the same problem, especially when being 'tested'. (If you want to 'revise' any of these areas, most of the current GCSE texts explain these ideas in a simple way.)

One project currently publishing detailed reports on primary schoolchildren's concepts in science is the SPACE project based at Liverpool University. SPACE (Science Process And Concept Exploration) has published research reports on fourteen areas of science concepts. The drawings you are asked to analyse in Stimulus Activity 3 were taken from these reports, which contain a wealth of detail about primary schoolchildren's science ideas.

But interpreting children's ideas is only the first step. The crucial decision to make is how to respond to the ideas children present to you. Science is a practical, investigative area of learning, so we need to decide what practical experiences are appropriate for a child in a given situation, as these will form the basis for the child's own reconstruction of his or her concepts. Stimulus Activity 3 should help you to choose the appropriate kind of follow-on experience; you need to try out the technique in your classroom science activities.

3. How Can We Find Out What Ideas Our Children Have?

It is a sign of the changing expectations of professional teachers that you are now expected to do this as part of the routine of teacher assessment in all core subjects. It is also a sign, however, of the growing recognition that professional development can best take place 'on the job' in your own classroom.

Teachers have always argued that they observe all the time in their classrooms, and in many ways this is true. But teachers who have had the opportunity to stand back and watch their own class learning have also usually realized that such observation is usually partial, unsystematic and often misleading in what it tells us.

A recent study of teachers learning through this kind of classroom observation and evaluation (Ashton *et al.* 1989) suggests that most of what teachers do is based on theories about classroom practice, theories which are often not made explicit. These theories in turn are based largely

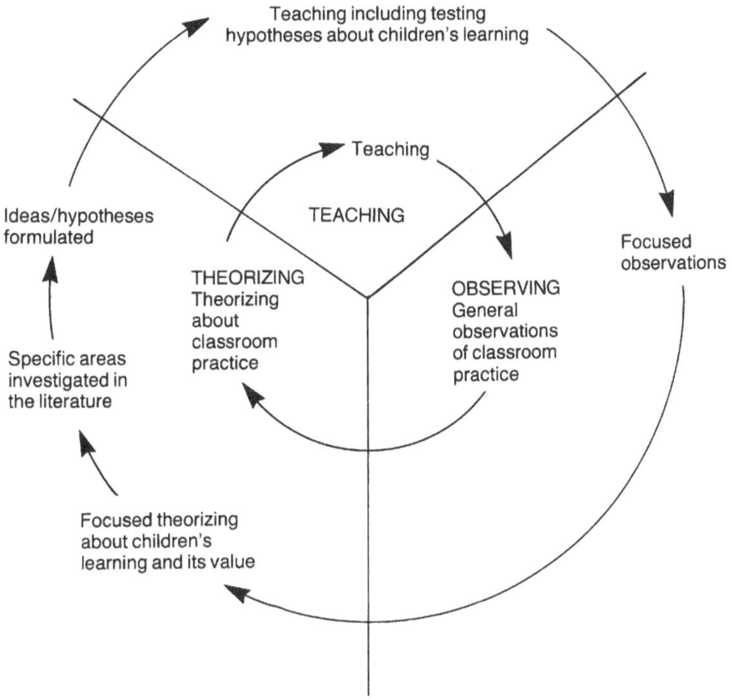

Figure 7.1 The observation–theorizing–teaching cycle

on experience arising out of general observations. One way of describing this process is shown in Figure 7.1.

The idea of practical theorizing, when applied to our current concern for identifying children's science ideas, means that you need:

1. to *focus* your observation on a specific skill or knowledge area being covered in the children's science work – this, incidentally, is also recommended in the SEAC 'Guide to Teacher Assessment in the Classroom' (1990) – and on a small number of specific children at a time: don't try to observe too much, for too long;
2. to *decide* how to record what you need, and *practise* recording in short bursts – do not be over-ambitious;
3. to *observe* what children are *actually doing* (and saying) in relation to your chosen focus;
4. to *discuss* what you think their ideas are and what they are learning with other teachers;
5. to *compare* your tentative conclusions with what others have found (e.g. in other classes, or in the SPACE research reports);
6. to *test* your ideas in a subsequent science session, either by observing other children on the same concept or the same children learning other concepts, etc.;
7. where possible, to have a colleague *observe you* whilst you are doing the above, to see what *you* were doing.

All this sounds incredibly daunting; another layer of things to think about whilst teaching! But you can be reassured by the evaluation by Ashton *et al.* already quoted, which confirms that many teachers around the country have not only trained themselves to work this way but also found it worth the effort, both for themselves and for their children. A crucial dimension is that of collaboration in the classroom; work with colleagues, student teachers, college tutors, advisers, parent helpers, or whoever is available to 'free' you to observe for short periods. As little as 10 minutes in a lesson can be extremely valuable. And as you will no doubt have noticed, the skills will be of equal value in carrying out essential Teacher Assessment in Science.

4. Summary

This Briefing Paper has emphasized the following points:

1. Children have ideas about science when they first come to school, based on their experiences. These ideas may be inconsistent with established scientific theories, but will seem logical and consistent to the child.
2. Doing science means testing these ideas against the evidence of their senses; so we have to provide children with practical experiences which will help them modify and improve their ideas.
3. Some common misconceptions are held by many children and adults, regardless of their cultural background. Being aware of these can help you select appropriate activities for your children.
4. To find out exactly what science ideas your children hold, there is no way but to observe them: watch and listen. You can train yourself to do this effectively, particularly if you do it in collaboration with others, and the skills will be of value in all aspects of your assessment work.

STIMULUS ACTIVITIES

Activity 1

Read the two brief extracts of discussion between teacher and child, and write down things that you might say in response, as teacher. You can write down as many responses as you want.

Example 1

ADULT (*showing photograph of Big Ben*): What is this a picture of?
FIRST CHILD: It's Big Ben.
SECOND CHILD: No it's not, Big Ben is a penguin.

Example 2

ADULT (*beating a drum with a stick*): What makes the sound?
CHILD: As you beat the drum sound echoes like I said and then comes up and seeps through the skin on top.

When you have considered carefully how you would respond, each member of your group should read their preferred responses to the others. One of you could produce a list of the different ideas generated, and you could then go on to consider which is the best way of responding, and why.

Activity 2

In the last activity, you probably became aware that the way you respond to a child's idea depends very much on your own understanding of the same idea or phenomenon, so it is useful to examine your own ideas about scientific concepts.

You can test out your own perceptions, first by trying to write down your explanations for each of the following phenomena, then by pairing up with a colleague and questioning each other's ideas in order to probe your understanding of the concept. Can you agree on what concepts are involved in each case?

1. When you fill a syringe with liquid, what makes the water go into the syringe?
2. Why does a pendulum swing?
3. What makes a piece of wood bob up if you hold it under water and then let it go?
4. How do you see things when the sun is not shining?
5. Why does the sound of a moving vehicle drop in pitch as it passes you?
6. Why do you feel cold when you come out of the sea or swimming pool?
7. How does a plant bend towards the light?

Activity 3

Here are three drawings by children, taken from the SPACE reports referred to in the Briefing Paper. In pairs, you should discuss and write down what ideas you think the children hold which have led them to make these drawings.

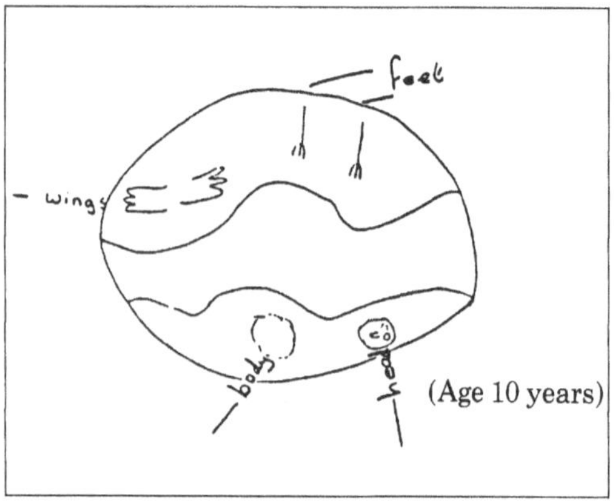

(Age 10 years)

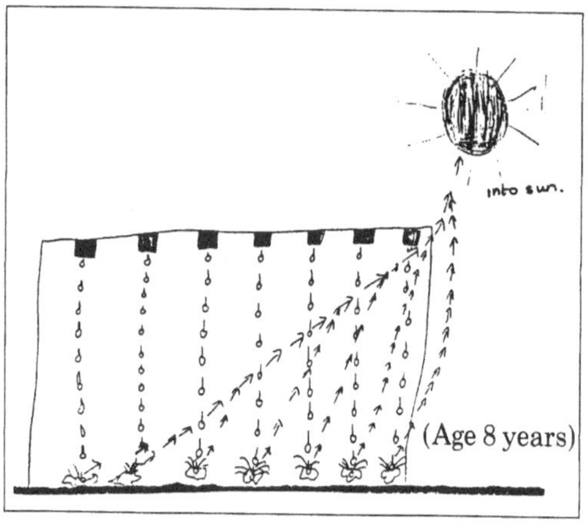

(Age 8 years)

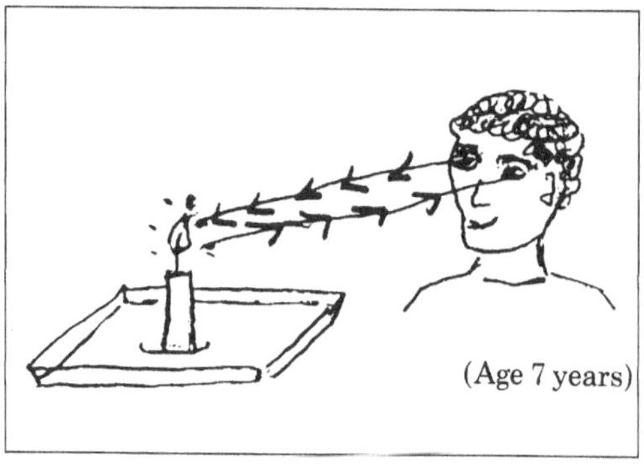

(Age 7 years)

The crucial decision to make again is how to respond to the ideas children present. In particular, with reference to AT1 of the National Curriculum, we need to decide what practical experiences are appropriate for a child in a given situation.

In the example below, choose which of the alternative practical activities you would want the child to do next, in order that she could modify her idea about the reason why a plant has a certain number of seeds.

> TEACHER: So a bigger plant has more seeds?
> CHILD: No, no, it depends on what sort of fruit you've got in there. If it's sour, if it ain't got no taste at all, it'll probably have a lot of seeds.

Alternative A: Give the child lots of examples of fruits with seeds to examine, so that he or she can make a more accurate generalization about the relationship between type of fruit and number of seeds.

Alternative B: Give the child sweet and sour examples of the same fruit (e.g. grapes, apples, plums), so that she or he will disprove their existing hypothesis.

Alternative C: Give the child big fruits with lots of seeds (e.g. melon, papaya, tomato), so that she or he will make the connection between fruit size and number of seeds.

In your group, you should then take turns to justify your chosen alternative.

Activity 4

As a group, observe a short video extract of children doing investigative science. You should try to apply the techniques described on p. 81 of the Briefing Paper in doing this. Make your notes on your observations individually, before sharing them with colleagues, in order to appreciate how different people 'see' the same situation in different ways.

Use the following checklist as a guide to your observations and analysis.

1. Which specific child am I observing?
2. What is the child actually doing/saying?
3. What science concepts are involved?
4. What has the child done/said in relation to that concept?
5. What is the idea that this child holds?
6. What would I want the child to do next?

CLASSROOM ACTIVITIES

Activity 1

On a day when your children are doing some practical scientific activity, pay particular attention to any examples of children revealing their ideas or misconceptions through what they do or say. At the end of the session, write down *one* example in as much detail as you can remember, along the following lines:

* what the child did or said that revealed his or her ideas
* how you interpreted this behaviour/statement
* how you responded at the time
* your subsequent thoughts on the situation

Keep a record of the above, to share with colleagues at your next discussion session.

Activity 2

The intention here is to examine your own explanations of science ideas to children.

On a chosen day when your class is doing some science work, pay particular attention to occasions when you explain an idea to a child or to the whole class.

At the end of the session, focus on *one* such instance during the day, and try to write down:

* What instigated the need for the explanation? Was it initiated deliberately by you? Was it a response to a question, a response to something observed?
* Why did you give an explanation rather than respond by asking a question, demonstrating or some other possible response?
* What language did you use? What scientific terminology did you use?
* Were you able to check if the child/class understood your explanation? How did you go about this?
* Looking back, are you happy with the explanation you gave? Are you confident now that it was adequate and helped develop the child(ren)'s ideas?
* If not, what do you need to do about it?

Activity 3

The intention of this activity is for you to try out a small-scale focused observation whilst you are teaching, and to try recording your observations at the same time. Ideally, you and your colleagues should agree on a common focus so that you will be able to compare findings.

Before beginning to observe, you should:

1. Decide which child or small group you are going to observe.
2. Identify a time slot of 10–15 minutes maximum when you will observe the child(ren), ideally when there is an opportunity to free yourself temporarily from involvement. Other adults in the classroom can make this possible; parents themselves are often keen to become involved in such observation.
3. Decide on *one* very specific focus for your observations. For example, you may wish to look at how the group organizes itself to undertake a science task; which child does what; who says what; who asks questions; how the group responds to intervention (from another child/student teacher/yourself); what they say to each other whilst recording results; how they explain their results. *Any one* of these is sufficient for a short observation.
4. Devise a way of recording what you observe, which does not involve making copious notes. You cannot look and write at the same time! Try a simple checklist or chart; having tried it and found it inappropriate, you will have learned how to make a better one for your next observation. The emphasis at this stage is on *learning to observe*, not on proving any particular point about children.
5. Tell the children what you are doing, in broad terms at least, or they may pester you all the time you are trying to observe, in order to find out!
6. Think about what you *expect* to find from your observations.

During your period of observation, try to resist the temptation to respond to children's demands, and try to stick pretty ruthlessly to the task you have planned. One teacher I know who was very successful at this used the simple device of a large badge with an eye on it. Whenever she was wearing the badge, the children knew not to interrupt!

After you have done your 10–15 minutes' observation, put down the checklist and forget it until a time when you are able to sit down and reflect on it. Do this on your own, before you share any ideas with colleagues.

Activity 4

Drawing together and finding words for what you have done and learned is an important stage in any professional development process, and needs to come before you discuss your ideas with colleagues. Some alternative ways of doing this are suggested below. You should try at least one of them.

1. Produce a display of children's work which provides evidence of their developing science ideas. Make captions to accompany the display, which explain what the work illustrates and the concepts the children are grappling with.
2. Arrange a time when a group of students (and their tutor) can help you to continue exploring children's science ideas in your classroom. Spend the first session setting out what you have done so far, and what questions you are still seeking answers to. Most training institutions will be very pleased to be invited to work alongside you in this way.
3. Plan your scheme of work for next term, taking account of what you have learned, making explicit reference to how this has come about.
4. Produce a short report on what you have learned for the head, or for the next parent/governors meeting, or for a widely read journal like *Primary Science Review.*
5. Discuss your ideas with a teacher from your neighbouring high school.

REVIEW SESSION

1. Share your example from Classroom Activity 1 of children's ideas or misconceptions, how you interpreted it, how you responded, and your subsequent thoughts on this.

2. Compare and discuss the various explanations each group member made in Classroom Activity 2, focusing on the questions in that activity.

3. Compare your findings from Classroom Activities 3 and 4 – small-scale, focused observation whilst teaching – using the questions listed there. You may well have found that, despite being ruthless and single-minded, you were not able to stick precisely to the observation you planned! Analyse exactly why this was the case:

* was it to do with defects in what you planned?
* were there unforeseeable circumstances?
* was there a classroom crisis? . . . and so on.

What can you learn from your and your colleagues' experiences to make your next observation work better?

4. This Unit has focused on three questions:

* How can we find out what our children's ideas are?
* What ideas do children in general commonly have about science concepts?
* How do we respond to these ideas?

In order to answer these questions, we have suggested ideas to do with practising classroom observation.

Make a list of all that you think you have learned in relation to these questions. Spend some time looking for and discussing patterns in what you have learned. In particular, look again at the Practical Theorizing Cycle in the Briefing Paper (p. 81). What ideas or theories has this Unit thrown up that you might test during your future teaching?

Make a list of those areas in which you didn't learn, didn't make much progress, or still aren't clear. You can then see which are common areas of concern. Why did few of you make any progress in these areas? Was it:

* because of some inadequacy in the Unit?
* lack of opportunity to try things out?
* results which didn't match what you expected?
* lack of confidence?
* not fully understanding what was intended?
* lack of science knowledge?
* fatigue?

As a group, discuss how you are going to tackle these areas and what you are going to do next.

5. You were asked to read relevant sections of some of the following publications referred to in this Unit. If sections were divided up amongst participants, report back now, explaining what each section was saying, and how it relates to your own observations.

References

Ashton, P. M. E., Henderson, E. and Peacock, A. (1989) *Teacher Education through Classroom Evaluation*, London: Routledge.
BBC Publications (1990) Primary Science INSET videos.
Driver, R., Guesne, E. and Tiberghien, A. (1985) *Children's Ideas in Science*, Milton Keynes: Open University Press.
ILEA Learning Resources Branch, Primary Science in Division 7, set of six videos, London: ILEA Learning Resources Centre.
Open University (1983) *Curriculum in Action*, Milton Keynes: Open University Press.
Schools Examinations and Assessment Council (1990) *A Guide to Teacher Assessment*, Pack A: Teacher Assessment in the Classroom, London: SEAC.
Space Research Reports (1990 onwards) Liverpool University Press; current titles include: *Growth, Sound, Electricity, Evaporation* and *Condensation, Light, Materials.*

UNIT 8 ACHIEVING PROGRESSION

LEADER'S GUIDELINES

This Unit aims to help teachers develop usable procedures for improving their ability to ensure progression. It focuses on three main aspects of progression – namely, Diagnosing, Planning and Matching. The Unit is aimed at all teachers; it focuses on procedures, rather than specific stages, so that work on the Unit can be shared between teachers of children of any age range. It also encourages tentative solutions to what is a difficult problem, on the grounds that final solutions are unlikely to be found. The nature of children's progression requires that it has to be attended to over an extended period of time; hence there needs to be a fairly lengthy period between starting the Unit and undertaking the Review Sessions. The Unit could usefully be undertaken simultaneously with other Units, such as Unit 5, 'Assessment', or Unit 7, 'Getting to Know Children's Ideas in Science'.

Briefing Paper The purpose of the paper is to present some of the key ideas associated with progression, and is therefore best distributed at the beginning of the Unit, and certainly before the Stimulus Activities are carried out.

Stimulus Activities Activity 1 will require about 45 minutes. Activities 2 and 3 will need about 35–45 minutes altogether to be tackled by the whole group.

1. Completing a Progression Chart. Participants will need copies of the chart on p. 95. The diamonds could be filled in as a whole-group exercise, or sub-groups could tackle one diamond each. Alternatively, groups of individuals could take the chart away and come to a Review Session with prepared suggestions.
2. Using the completed Progression Chart to work out progression requirements for two hypothetical case studies (Fiona and Raoul).
3. Group decision making about progression issues such as how many activities to offer at a given time, grouping, assessment implications, and so on.

Classroom Activities Each of these four activities involves teachers observing an individual child or a group systematically over a period of time. Allow plenty of time (up to half a term) for teachers to complete some, if not all, of the activities. Ensure that, in making their observations, participants record evidence to bring to one or more Review Sessions.

1. Producing a 'Science Profile' of a child.
2. Children reporting back on their findings and problems, and suggesting ways forward.
3. Allowing children to progress from one skill level to another by means of a small circus of science activities.
4. Experimenting with group sizes during a practical science investigation.

Review Sessions

Because all the classroom activities involve the observation of children over a period of time, it may be better to hold a number of extensive Review Sessions for teachers to report back on the various Classroom Activities. We suggest the following:

1. When a number of teachers have produced Science Profiles of individual children (Classroom Activity 1), they have a session at which to review each of these profiles. (It might also be useful to circulate the Science Profiles of individual children amongst participants prior to the Review Session, for comparison. You may need to make them anonymous first.) The focus should be on the process of helping a teacher to produce evidence for the conclusions in the profile. The point is not to make definitive decisions about a particular child, but to help develop procedures for diagnosing children's levels of attainment in science. This would need up to an hour to consider several cases effectively.
2. This session should be used to review the group work undertaken in Classroom Activities 2, 3 and 4. Again, the focus should be on helping one another to establish procedures, to present 'progress reports' and interim plans, to be reviewed again at later stages.
3. It would be helpful to have an open-ended Review Session which simply starts 'My current problem is . . .', and invite others to help in brainstorming/trouble shooting in relation to matters of achieving progress. It is important to encourage tentative suggestions and to encourage others to test them.

The outcome of these sessions should not be one which leaves teachers feeling that it is all too difficult, but one which allows them to go away with another idea to try out which might help.

You might wish to consider sharing a Review Session with a review on Unit 5, 'Assessment', or Unit 7, 'Getting to Know Children's Ideas in Science'.

BRIEFING PAPER

Progression in Learning Science: Some of the Difficulties

Progression is one of those 'big ideas' in education which is fairly straightforward to understand, but pretty difficult to achieve. We all accept that children must move forward in their learning on to new, more challenging ideas and skills; and most teachers have many ideas and activities for children to undertake. So what is it that prevents children from progressing naturally in their learning?

Evidence from HMI and research studies has suggested that activities are less well matched to children's requirements in science than in other subjects, and that teachers have difficulty, in busy classrooms, in diagnosing children's immediate learning requirements. In particular, teachers find it relatively easy to assess when work is too hard for a child, but often don't notice when work is too easy. In addition, the authors of the science National Curriculum initially suggested that knowledge about children's science learning is not yet adequate for the purpose of setting out a clear progression through levels, since process skill learning is very context-dependent – progress in observation skills when dealing with living things, for example, might not transfer to progress on observation skills relevant to accurate measurement.

What does progression involve, then, for primary children doing work in science? In theory, it means three things:

1. *diagnosing* the level of attainment of each child in relation to the activity they are undertaking, bearing in mind both the skills dimension (AT1) and the Knowledge and Understanding (ATs 2–4);
2. *planning* activities which allow children to move from one level to the next, or which consolidate the level which they have just reached;
3. *matching* the child's immediate needs with one of the planned activities in order that he or she will be able to progress.

The next section therefore describes one example of how a teacher might actually deal with this in practice.

Progression Within a Science Topic: a Practical Example

Let's imagine that you are planning a topic on 'transportation' with a Year 5 class (9–10-year-olds). You will need to make sure that your activities deal with AT1; and in practice, you may also need to ensure that you cover all levels from 2 to 5, if the class has a wide spread of attainment. But in order to keep this example manageable, you might concentrate on only one aspect of AT1: namely, 'fair testing'. You also need to make sure you deal with one or more of the Knowledge and Understanding Attainment Targets: in this case, perhaps Materials (AT3) and Forces (AT4).

You also need to remember that you have to combine both the knowledge and skills elements in your activities. A useful way to remember this is as follows:

CONTENT	=	CONCEPT	+	CONTEXT
(i.e. description of the activity)		(what the learning will focus on)		(the medium through which it will be learned)

The *concept* can be from the knowledge ATs (e.g. 'the factors that cause objects to float or sink') or a skill ('constructing a fair test'). Either way, you have to put it in context – which means, for instance, that:

* to learn about factors which cause objects to float and sink (knowledge concept), they might carry out an observation of what happens when different objects are placed in water (skill context)
* to learn to construct a fair test (skill concept), they might need to try out different materials for making boats (knowledge context)

The Transportation Topic: What Are the Relevant Skills and Knowledge?

If you look at the Statements of Attainment related to 'fair testing' in AT1, you can see that progression within the skill is characterized as follows:

level 1	no mention
level 2	no mention
level 3	distinguish between a fair and an unfair test
level 4	construct fair tests and control variables
level 5	identify and manipulate relevant dependent and independent variables, choosing appropriate ranges and values

The relevant knowledge obviously depends a great deal on your particular interests, your school environment, what other classes are doing in your year and in other years, and so on. For the sake of this example, let's imagine you are doing 'Water Transport'. If you look at AT3 (Materials), it is possible to identify the following options:

level 2
* classifying (boat-building) materials according to their characteristics

level 3
* distinguishing man-made from natural (boat-building) materials

level 4
* comparing these materials for strength, hardness, effect of water
* relating the characteristics of materials to their use (in the boat itself: hull, mast, deck, sails, etc.)

level 5
* not relevant to topic

Looking at the 'Forces' example (AT4), another set of possibilities arises, as follows:

level 2
* understanding that pushes and pulls can make (a boat) start or stop

level 3
* understanding the factors that make things float or sink

level 4
* understanding that movement (of a boat) depends on the size and direction of the force acting (e.g. in sailboats)
* knowing how to measure weight (in newtons) and its effect (in, e.g., loading ships)

level 5
* understanding that when an object is not moving, or moving at a steady speed, balanced forces are acting on it
* testing the strength (of a boat or part of its superstructure – e.g. its loading davits, sail, deck)
* measuring force, distance and time in relation to the movement (of a boat across water)

Progression involves *diagnosing, planning* and *matching.* Let's assume, for now, that you have assessed children's levels and diagnosed their difficulties (see Units 5 and 7 for activities relating to this). The *planning* stage requires you to think up activities which combine the skill and knowledge needs as set out above, and which build on previous successfully completed activities. A way of doing this is through a Progression Chart, which you can construct in Stimulus Activity 1. What this chart helps you to do, when completed, is to see what the next logical activity is for a child at any given stage. Stimulus Activity 2 asks you to use the chart to plan the next activity for two hypothetical children, Fiona and Raoul.

Progression in Group Practical Work

One way to facilitate progression might involve having several science activities in a topic set up at 'stations' in the classroom, so that children could move on to a relevant activity without disruption to the teacher or the rest of the class.

However, nobody wishes to suggest that children should undertake activities individually, since many of the planning and communication skills can only be acquired when working collaboratively in groups. So how do we deal with the situation described above where Fiona and Raoul, who started out in the same group doing the same activity, may need to progress on to different activities for their next investigation?

Stimulus Activity 3 asks you to consider a suitable menu of activities and the case for different groupings, time spent on activities and so on. Following on from there, Classroom Activity 2 requires you to try out a group feedback procedure which not only encourages children to report on their own progress but also provides a valuable way of diagnosing their particular difficulties with an activity. Both of these lead to the involvement of children in helping determine what is appropriate for their own progression.

Differentiating Activities

An alternative approach might be to keep the class doing the 'same' activity, but to differentiate that activity according to the level appropriate to individual children. An example of this approach is provided in the National Curriculum Non-Statutory Guidance in Science, pp. C8–C13 (GB, DES 1989). In this example, all the children are undertaking an activity to construct a windmill; however, the activity is presented in four different ways, each of which makes different demands on the children. For example, one approach gives clear instructions to the children about how to make the windmill, and only asks them to draw or paint it as a way of recording their activity; in other words, the approach is almost entirely at level 1, in both skill and knowledge terms. Another approach, on the other hand, requires children to control certain variables in the construction of more than one windmill, in order to compare their efficiency, and to present findings as a graph. This approach is largely aimed at level 3, in both skill and knowledge terms.

Transferring this example to our topic on 'Water Transport', it is easy to see how the activity involving designing and testing two boats with different hull shapes could be modified to render it appropriate at lower or higher levels, as required. For example, providing clear instructions on how to design and make the boats would bring the activity down to a lower skill level; whilst introducing other variables besides the hull shape (such as construction material) would increase the knowledge level.

In such a way, any of the activities you have created in your Progression Chart can be modified so that, within reason, children who have been collaborating in a group on earlier activities could stay together and undertake further differentiated investigative work.

Conclusion

At the beginning of this paper, you were reminded that progression is a deceptively simple idea in theory, whilst being extremely difficult to achieve in practice. You should not feel inadequate or overwhelmed if you only occasionally succeed in ensuring a smooth progression for the

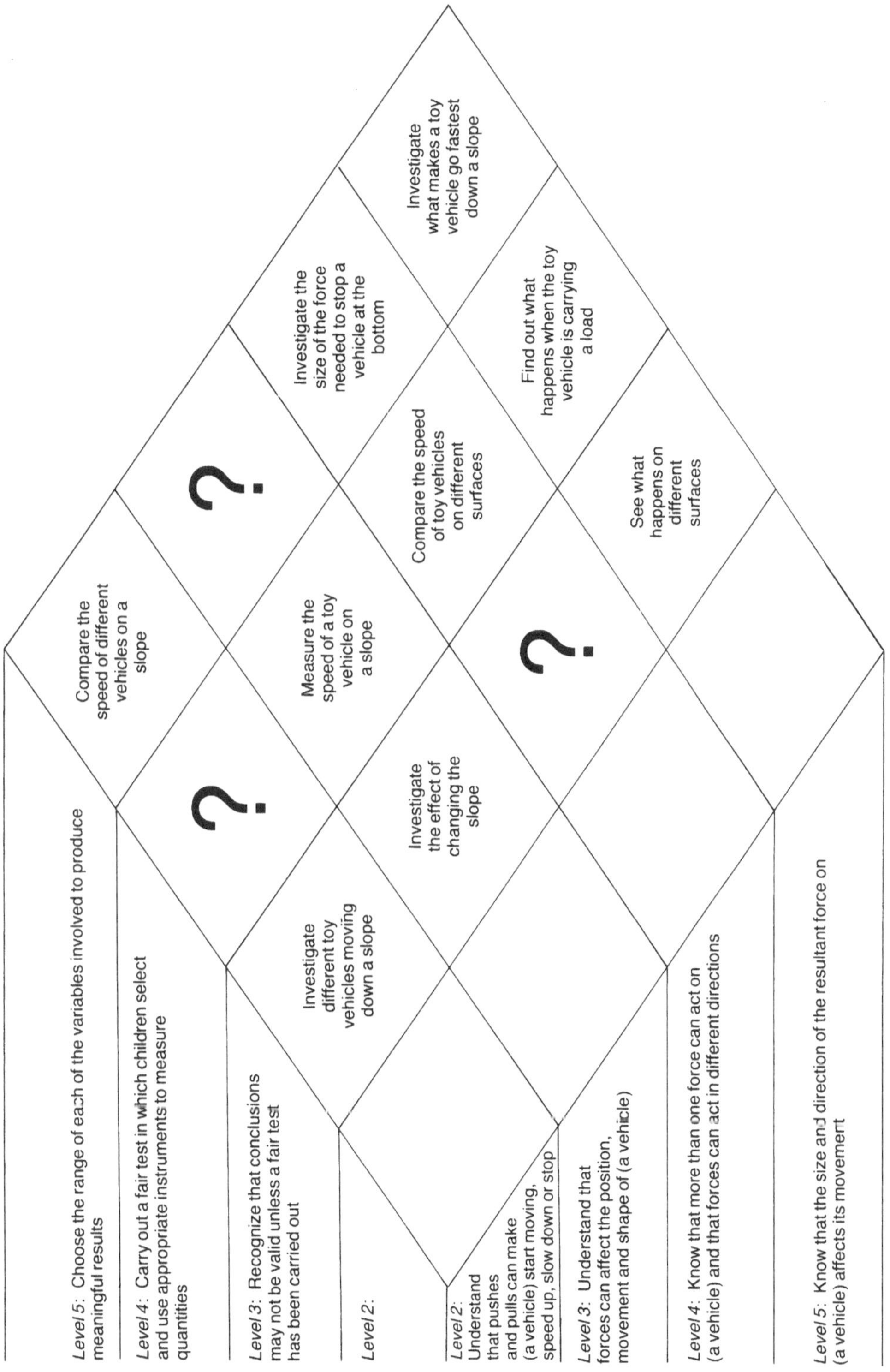

Figure 8.1 Progression chart for topic on 'transportation'

children in your class. It is important to remember that the authors of the National Curriculum science document, in their interim and final proposals, drew attention to the 'provisional nature' of their conclusions, and admitted that they could not confidently place the investigative skills of science into a ten-level hierarchy which had progression built in. Perhaps when several generations of children have worked through the document, and teachers have solved all the problems of diagnostic assessment, we will be able to assure future colleagues that we know how to ensure skills progression; but at present we have to accept that it is still an inexact science!

There is one thing, however, about which the Science Working Group was sure, which is expressed in para. 2.14 of their final proposals, as follows:

> *As a result of learning Science, we expect children to make progress in their use of investigative skills. Such skills, however, do not progress in isolation from the subject matter or ideas that are being used.*

Hence the importance, in attempting to plan for progression, of keeping both skills and knowledge attainment targets in view at the same time. One way to do this is to make sure that topic planning begins with the Programmes of Study, since these encompass both knowledge and skills. And from then onwards, a planning device such as the Progression Chart (p. 95) is helpful in making sure that neither one dimension nor the other is lost sight of. Finally, it is also essential that teacher assessment takes into account both the skill and knowledge components of children's investigative work. This dimension is dealt with in Unit 5.

Reference

Department of Education and Science (DES) (1989) *Science in the National Curriculum*, London: HMSO.

STIMULUS ACTIVITIES

Activity 1

In the Briefing Paper for this Unit, some of the skills and knowledge relevant to levels 2–5 for a topic were set out, by quoting from the relevant National Curriculum Statements of Attainment. In this activity you are asked, with help, to combine these into suggested activities in order to complete a Progression Chart. In true *Blue Peter* style, some of the chart has been done for you earlier, to provide examples of what is needed.

In Figure 8.1, for example, the 'fair test' statements are on the upper arm of the diagram, whilst statements about 'forces' are on the lower arm. The diamonds in the grid each contain a suggested activity, which attempts to combine the skill and knowledge statements at the relevant levels. For example, the activity 'Investigating different toy vehicles moving down the slope' combines the skill of recognizing the need for a fair test (level 3) with the knowledge that push and pull forces can make a vehicle move (level 2). Some diamonds are empty because there is no relevant skill statement at level 2; others (indicated by a question mark) have been deliberately left uncompleted, as an exercise for you. As a group, therefore, you should brainstorm ideas for activities which would be appropriate to the skill and knowledge statements at the relevant levels.

Activity 2

Progression, in terms of the chart you completed in Activity 1, involves children moving ultimately from *left to right* across the chart. However, progression for any individual child might not be a straight line. For example, you might notice in your assessment that a child has mastered the knowledge element of an activity, but not the skill element. Therefore it would be appropriate for that child to progress to another activity at the *same skill level*, but at a *higher knowledge level*.

Here is a hypothetical case. Use the chart to decide what would be the next appropriate activity for the child.

Fiona has just completed an activity in which she and her group were timing how long it took a toy truck to run down a slope, using a stopwatch. They could all explain that there was a force (gravity) 'pulling' the truck down the slope, and that if they made the slope steeper, the truck went faster. However, the teacher noticed that when she measured the time taken by the truck, Fiona herself did not always make sure that the truck started from the same place on the slope, and that the person doing the timing was not always the same.

Activity 3

In planning progression in science within a topic such as Transportation, you need to make some key decisions at an early stage about the way you will organize the children's work during the topic.

Look again at the kinds of activities in the Progression Chart you completed, and then as a group make decisions about each of the following:

1. How many such activities would you expect a child to complete in the course of the topic?
2. How many (and which) would be 'on offer' to the class at any given time?
3. What are the arguments for and against children doing these activities singly, in pairs, in small groups, and so on? Is it appropriate for children to work in different groups at different times?
4. What are the implications of your above decisions for teacher assessment?

CLASSROOM ACTIVITIES

Activity 1

Choose one child in your class and, over a period of two or three science activities, produce a Science Profile of that child, which attempts to answer the following questions:

1. What ideas does she or he have about the science issues we are concerned with?
2. What skills has he or she demonstrated during the activities, and at what levels?
3. What new knowledge has she or he acquired during the activities, and at what level is this learning?
4. What is the role he or she takes during group practical work?

Try to record evidence relating to the points you make above. You will need to return to this profile in the Review Session.

Activity 2

This activity should be undertaken with your whole class *during* the topic, but *not* at the end of it.

The activity is a simplified version of a 'research colloquium' where teams of researchers engaged on a similar problem come together to share their findings and problems, and suggest ways forward. Each of the groups in your class that has been undertaking science investigations should be asked in advance to prepare to talk to the rest of the class for 5–10 minutes, using the following agenda to guide their presentation:

* what is the problem we have been investigating?
* what ideas did we try out?
* how did we go about it?
* what problems did we come up against?
* what have we found out so far?
* what stage are we at right now?
* what is our strategy for the next step of our investigation?

Note that all the groups need not report at the same time; it is most fruitful if a group reports at a crucial stage in their investigation, or a turning point in their understanding. It is probably a good idea with older classes to let the groups themselves decide when they are ready to lead a colloquium. Let the rest of the class cross-question them and offer suggestions. If this strategy seems over-ambitious, rest assured: children take to it very readily, with practice, and I have seen it working effectively in schools in the inner city and in rural Africa.

You as the teacher do not need to know the answers to their problems: a problem is a starting point for an investigation. For example, if a group says, 'Our problem is, we don't know how to measure the push we gave to our boat', your best response might be (to the class as a whole), 'OK, what suggestions have we got about measuring push? What instruments might we use?'. They are then beginning to design and try out ways to solve the problem. And that's science!

Activity 3

With reference to the Progression Chart you completed in Stimulus Activity 1, set up a small circus of science activities which allow children to progress from one *skill* level to another. Observe what happens as groups of children proceed through the activities over a period of time, and note the following:

* which individuals are clearly progressing in their skill attainment?
* which are clearly not progressing?
* which individuals are you not sure about?

You will need to return to this in the Review Session.

Activity 4

During a session when the whole class is working at a practical science investigation, deliberately put children into groups of different sizes, from one to as many as five per group. Then note the following:

* what different ways of working did you observe in the different-sized groups?
* in what size of group did all the children seem to learn best?
* in what size of group was it easiest/most difficult to assess their learning?

REVIEW SESSION

Since all the Classroom Activities involve observing children over a period of time, it is suggested that more than one session be held, to include the following.

1. Hold a session to review the Science Profiles undertaken in Classroom Activity 1. Focus on the following:

* What evidence do you have for your conclusions in the profile?
* How could you develop the procedures for diagnosing children's levels of attainment in science?

2. Review the groupwork undertaken in Classroom Activities 2, 3 and 4. Focus on establishing procedures, e.g:

* What strategies can the group suggest to overcome the problem that, even after several activities, there are always some children whom we are 'not sure' about, in terms of their learning/progression?

In relation to Classroom Activity 4 on group size, after sharing your findings, address the following questions:

* From the point of view of helping individuals progress, is there an optimum group size for a science activity?
* Does the optimum size vary from individual to individual?
* In what ways does working in a group help or hinder an individual's progress?
* If group size is or is not important, what does this imply for our organization of science activities in the classroom?

3. In relation to achieving progression in science, what might you come up with for the following?

'My current problem is . . .'

Brainstorm solutions to this and any problems that participants have encountered relating to matters of achieving progression. Feel free to make tentative suggestions and test them out; after all, that is what science is about! Do not feel you have to come up with once-and-for-all solutions.

UNIT 9

WIDENING THE RANGE OF CHILDREN'S REPORTING STRATEGIES IN SCIENCE

LEADER'S GUIDELINES

The purpose of this Unit is to encourage participants to gain a wider understanding of the various ways in which children can communicate their work in science.

Briefing Paper

You should distribute this to participants after the Stimulus Activities, as they will require it to help them to undertake the Classroom Activities.

Stimulus Activities

There are four activities in this session, and you should allow about 1½ hours.

The four activities all involve participants in experimenting with different ways of reporting:

1. Newsletter report on getting rid of moles.
2. Trying to convince the family teenagers to use less detergent in the washing machine.
3. Designing some user-friendly instructions to operate a video-recorder.
4. Designing a more appropriate and effective way for a child to record her science work.

Participants should be asked to work in pairs or larger groups, and to do *one* of Activities 1, 2 or 3 and then Activity 4 – for which you will need to provide copies of Stimulus Activity 1, Unit 5. In this way they will write about science 'at their own level' as well as thinking about ways of encouraging children to report their work. They should then be asked to talk about their task to the whole group.

Classroom Activities

Having read the Briefing Paper, participants should identify two or three different ways in which the children in their class can record their work for each of the next three science activities that they undertake. Allow at least two to three weeks for this work to be carried out in the classroom.

Review Session

You will need a half day's work or two sessions of 1½ hours each for participants to decide on the list of criteria for evaluating the appropriateness of the various methods of recording. It would be helpful for all the participants together to decide on other criteria which can be added to the list. Then they can work in smaller groups and evaluate the ways in which the children have recorded their science.

BRIEFING PAPER

The aim of this Unit is to give you the chance to explore a range of different ways in which children can be encouraged to communicate their work in science.

Real Scientists

It is worth starting by considering how 'real' scientists make records, for whom they make them and why they do it. Frequently, they are communicating to themselves. When they are planning and carrying out investigations, they will often use a notebook to jot down ideas and observations. These can make fascinating reading – as long as you can understand their own particular form of shorthand! They also use their notebooks to help them to reflect and make sense of their findings – or just to work things out.

Scientists need to communicate with others who are working in the same field. They may do this informally by talking to each other face to face or over the telephone, by writing a letter, by taking photographs or sharing a computer print out.

Once they are at the stage when they are ready to describe their ideas and findings in a more formal way, then a scientific paper is written and submitted to the appropriate magazine or journal for publication. This kind of communication usually follows an agreed format and has given rise to the headings traditionally used in secondary school science: 'Aim, Method, Result and Conclusion'.

Some scientists believe that they also have another task – one that is arguably more important and more difficult – to inform and excite people who have only a limited understanding of science. Drawings, photographs and films are often used to provide visual images, but it is the style and the use of language that are the most powerful tools for this purpose.

Children as Scientists

One of the tasks for any primary teacher is to enable the children in his or her class to undertake activities that are not only intellectually challenging but that are also authentic. In the case of science this means that the children require situations where they can ask questions, make predictions, formulate hypotheses, plan and carry out investigations and, of course, record and report their work and findings. Just as we want them to learn how to plan investigations for themselves, so we also want them to be able to make their own decisions about the most appropriate way of communicating and recording their findings. First, they need to be aware of, and to have experienced, the wide range of strategies that are available. To do this you need to help them to answer the following questions.

* *Who* will they be communicating to? themselves? Other children in the class? other classes? their parents? you, their teacher? the headteacher? someone whom they have not met?
* *Why* will they be recording what they think, do or find out? Is it to help

them to plan their investigation? To enable them to have a record of what happened so that they can make sense of it later? Or to make a comparison with other findings? Is it so that others can know what they have been doing?

Bearing these two – *who*? and *why*? – in mind, then

* *What* needs to be recorded or communicated?
* *How* should it be done? through drawings, diagrams, formal writing or a poem? using a camera, a tape-recorder, a computer? making a model or a wall display? through drama or dance?
* *When* should they do it? Before they start the investigation? As they work through the task? At the end? Or a mixture of all three?

What Does the National Curriculum Suggest?

The general introduction to each of the Key Stages includes a section on 'Communication'. It is worth reading these in full, but the following extracts will remind you of some of the ideas put forward for Key Stages 1 and 2.

Key Stage 1

> *They should [also] have the opportunity to express their findings and ideas to other pupils and their teacher, orally and through drawings, charts, models, actions and writing.*

Key Stage 2

> *Pupils should have opportunities to continue to develop and use communication skills in presenting their ideas and in reporting their work to audiences, including pupils, teachers, parents and other adults. In giving an account, orally or in writing, they should be encouraged to present information in an ordered manner. They should be introduced to the conventions used in diagrams, tables, charts, graphs, symbols and models. . . . Pupils should use computers to store, retrieve and present their work.* (DES 1991)

Some of the Strategies that are Available

Talking to others This may be the most appropriate way for very young children to communicate their ideas to you and to others in the class. Older children can be encouraged to give individual and group reports based on their work, explaining why they carried out an investigation in the way they did, as well as what they found out. A parent's evening or science fair can provide a good excuse for making audio- or video-recordings of this kind of report.

Drawings and diagrams Again, this is one of the major ways in which we expect children in Key Stage 1 to record their work. You can encourage them to add a sentence or two to their drawing, or to annotate it and thus identify its various parts.

Long explanations or descriptions are usually best avoided in science reports. They tend to detract from the main issues and can be replaced more effectively by drawings and diagrams.

A series of drawings, perhaps using the idea from a comic strip, can be used to describe the whole investigation. By suggesting this idea to children, you are encouraging them to think about how they can report their work as a sequence of events.

Making a detailed drawing of an animal or plant, from life, can enable a child to gain a fuller understanding of this living thing, as he or she has to observe carefully, and to look for things that are significant but not

Figure 9.1

immediately obvious; have a look at the Australian leaf insect in Figure 9.1.

Writing

Using notebooks Children need to be encouraged to have their rough notebooks or 'jotters' beside them during a science activity. These can be used in the same way in which scientists use their personal notebooks. The children can make drawings or write down ideas during the planning stages of an investigation, and they can have several attempts at recording their observations before they 'go public'. Children's preliminary ideas about what they have 'found out' or what patterns they can see in the data can often be resolved through jotting down their thoughts – without having to think too much about spelling or presentation!

A formal report There are several ways in which children can be encouraged to write about their science activities in an ordered manner, with an appropriate sequence of happenings and ideas. One of them, the comic strip, has already been suggested.

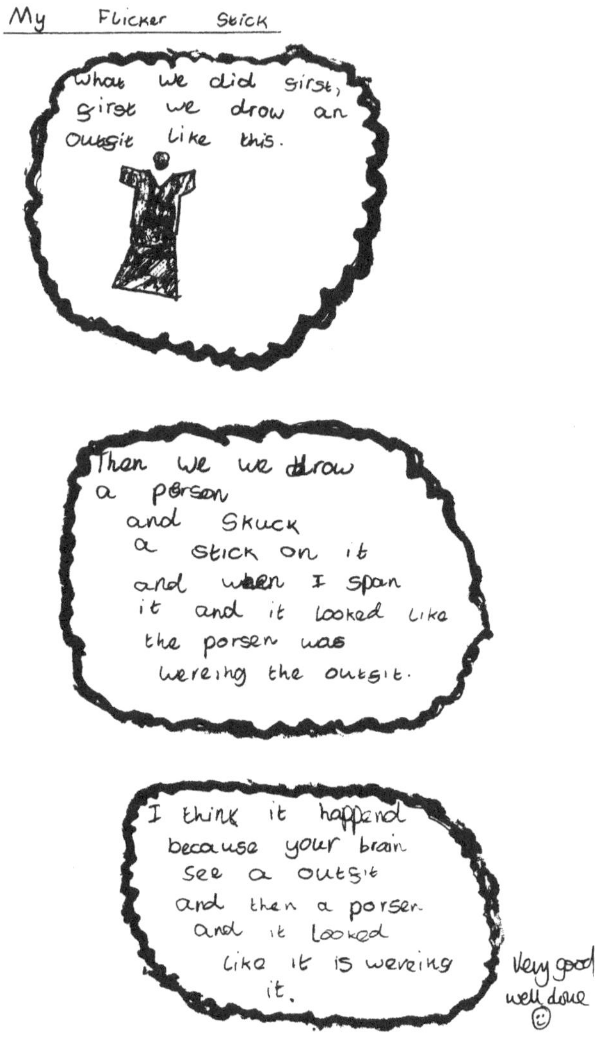

Figure 9.2

Another way is to divide the task into several well-defined parts. Figures 9.2 and 9.3 show how Jane and Alex have described how they made their flicker sticks and what they saw when they twisted them. Their teacher had suggested that they should use the heading 'What we did' for the first section. The children then answered the questions 'What happened?' and 'Why do you think it happened?', thus creating two further sections. Writing the phrases within 'balloons' can serve several purposes. It limits the amount of space available for the children who have difficulty in writing many words, so that what they do achieve looks complete. It also encourages an older or more able child to think about what to include and to leave out the more irrelevant information. These two descriptions were part of a class display which included the 'flicker sticks' themselves.

Later on in the school, when these children will be expected to make an

Figure 9.3

ordered record of a science investigation, they will be given the following headings as a framework for their report:

* what we wanted to find out
* what we did
* what happened
* what we think now

As a lead up to this style of recording, teachers may use several strategies. For instance, they can get the class to help them to use these headings to describe their work, leaving one or more sections for the children to complete.

Also, the teacher can ask them to focus on a particular aspect of 'what we did' by using one or more sub-headings, like these:

* the things we used
* a drawing of our experiment
* how we made it fair
* what we expected to happen
* what we measured
* a table/graph of our results
* some of the problems we had

Other ways There are many alternatives to the formal way of describing a scientific investigation, and they can be used to widen children's experience of different writing styles. They also help to avoid the situation where, by insisting that all science has to be 'written up', you dull the interest and excitement engendered by a practical investigation. Some examples are given below – you may be able to add others:

posters
giving directions or instructions
newspaper articles
letters
poems
stories
scripts for a drama production
using symbols, e.g. bulbs, batteries and electric circuits
diaries
devising a game – with rules

Matthew and his friends have been thinking about the purpose of a vertebrate skull during their study of evolution. You can read his ideas in the extract below.

> *WHAT IS THE SKULL'S JOB?*
> *A skull is a crash helmet for your brain. Without a skull, when you were asleep and rolled over, your face would all twist up and go funny. You'd be able to squeeze your head down the plug-hole, and you'd be able to squash it up into the size and shape of an economy-sized tube of toothpaste. Instead of a skull you could have a kind of structure on the outside of your head. There are two other alternatives. Your head could be filled with air. Then it would look like a big balloon. The only problem is, you might get a splinter or cut in your head, and then your head would go pop, and bits of brain and blood would fly everywhere. Or your head could be filled with water, and then if you banged it against the wall, it would all go wavy and wobbly like a droplet of water under a microscope.*
> *by Matthew*

Using maths Many science investigations involve children in sorting, counting and taking measurements. These can sometimes be represented by a drawing, but it is usually more appropriate to construct a table of results. Often a graph can then be made, enabling the children to make comparisons or to identify what happened over time.

Using information technology Concept keyboards are a useful tool for helping young or less able children to record their work in science.

Computer databases and spread sheets can be used to record observations and results, which can then be analysed in a variety of ways.

Data-logging sensors can help to investigate and record different aspects of the environment – for example, the temperature of the school playground at different times of the day.

Reference

Department of Education and Science (DES) (1991) *Science in the National Curriculum*, London: HMSO.

Acknowledgement The authors wish to thank the children and staff of John Stocker Middle School, Exeter and Park Junior School, Dartington for their contributions to this section.

STIMULUS ACTIVITIES

Activity 1

As a result of good luck and hard work, the lawn in your front garden looks like green velvet! One morning you notice that its professional appearance is marred by several molehills. Living in a country district, you are not short of advice from others, and are told to:

1. bury a beer bottle upright in each of several molehills with several centimetres of the neck above the surface of the soil;
2. stick a child's windmill into a molehill so that it catches the breeze;
3. bury two or three mothballs in several molehills, trying to get them down to the run.

When you attend a meeting of the local gardening association, someone suggests that you find the best way of encouraging the moles to move on and that you write a report of your investigations for the club's newsletter.

First, you need to decide what you know already about moles – and what you need to find out. Then:

* predict which method is going to be the most effective
* design your investigation
* write your report for the newsletter (for the purpose of this activity you will have to decide now which will turn out to be the best method!)

Activity 2

You have recently bought a liquid detergent to use in your washing machine. You realize that you need to decant some into the 'wash-ball' and have read the instructions carefully. These say:

	Lightly soiled	*Moderately soiled*	*Heavily soiled*
Soft water	60 ml	80 ml	140 ml
Hard water	80 ml	100 ml	160 ml

Living in a district where the water is particularly soft, you wonder whether the detergent would be just as effective if you used less than the recommended amount. And what about the water temperature? Could you economize here, or would that mean using more detergent?

Decide how you would answer these questions and how you would record your results in order to convince the teenagers in the family to use less detergent.

Activity 3

Your elderly mother has recently bought a new TV set and video-recorder. Whilst appreciating the facility to record her favourite programmes, she finds the remote control unit difficult to understand.

Design some friendly instructions that can be written on one side of a piece of A4 card and which will enable her to record a programme when she is not at home.

Activity 4

Read Stimulus Activity 1 from Unit 5, 'Assessing Children's Learning in Science'. Can you design a more appropriate and effective way for Anna to record her work?

CLASSROOM ACTIVITIES

Use the ideas in the Briefing Paper to identify two or three different ways in which the children in your class can record their work for each of the next three science activities that they undertake. Obviously you need to take into account their previous experience when making these decisions, but the purpose of this task is for you to be able to compare different ways of recording the *same* investigation.

Bring a selection of this work to the Review Session.

REVIEW SESSION

1. Within your group decide upon the criteria that you are going to use in order to evaluate the appropriateness of each of these various methods of recording. You might like to add to this list:

* Has it encouraged the children to explore their own ideas and to reflect upon their work?
* Has it helped them to make connections between what they wanted to find out and what happened?
* Does it enable you to learn more about their understanding of the science content and skills that were involved in the activity?

2. Taking one activity at a time, and using these criteria and others, decide which of the ways of recording was the most appropriate for each activity.

UNIT 10

CREATING A DIALOGUE WITH PARENTS

LEADER'S GUIDELINES

This Unit focuses on three main aspects of teachers' links with parents over science: namely, informing parents; listening to parents; and involving parents in science activity.

It would be preferable if Unit 10 were undertaken later rather than sooner, and not before Unit 5, 'Assessment'. However, since the Classroom Activities may need to be spread over a long period, you may wish to introduce the Unit and 'flag' these activities earlier.

Briefing Paper

The Briefing Paper should be given to participants to read after they have undertaken Stimulus Activities 1 and 2. They will need to have read the paper before they undertake Stimulus Activity 4.

Stimulus Activities

The four activities should be spread over two sessions of 1½ to 2 hours each.

1. What do parents already know? Using a 'Diamond Sort' technique to prioritize participants' perceptions of parents' knowledge of their children's science work.
2. Ways of communicating with parents – investigating all the possible channels of communication with parents before making decisions about the most appropriate.
3. Simulated communication exercises – situations resulting from inadequate communication between home and school.
4. Choosing the right strategy – comparing different modes of communication in practice.

1 and 2. These can be undertaken in the first session, with a break in between. Activity 1 will probably raise more questions than can be dealt with in the time available; before moving on, you should discuss with participants how and when to deal with the questions remaining. Participants' notes from both sessions, including their 'Diamond Sort', the Communicating Chart in Activity 2 and any notes, should be brought to the Review Session. The Briefing Paper should be distributed at the end of Activity 2.

3 and 4. These can also be undertaken in one session with a break between. Activity 3 uses simulation material (p. 122), and this should be distributed at the beginning of the session with time allowed to read it; or, if preferred, you might select and allocate examples at the end of the previous session, so that participants can think through their responses before coming to this session. In this way, more of the issues could be dealt with.

Prior to Activity 4, divide the participants into five groups. Assign to each group one of the modes of communication listed below:

1. a letter to parents;
2. a science fair or display of children's science work;
3. a parents' evening or meeting;
4. outside assistance (local media, sponsors, higher education institutions);
5. a combination of the above.

The drafts produced by groups and their balance sheets of advantages and disadvantages should be brought to the Review Session. It would, however, be helpful to provide a brief opportunity to look at one another's draft ideas before the end of the session.

Classroom Activities

These are not strictly 'classroom' activities since they do not involve children, but each will need time to implement. Given that displays, parents' meetings and so on do not take place frequently, you may need to allow at least half a term, or maybe a whole term, before organizing a Review Session to look back over the activities.

1. This involves systematic monitoring of one way of communicating with parents, such as sending out reports, organizing meetings or a science week. The main purpose is to collaborate in monitoring, so participants working in the same school/year group should work together where possible. You should agree a date with the whole group for the Review Session, and work to this as a deadline. You will need to do this at the end of Stimulus Activity 4.
2. This involves planning and evaluating discussions with individual parents at parents' meetings. The whole group will therefore need to be consulted on an appropriate time for the Review Session. It might be best to sort this out at the very beginning of the Unit, so that meetings are not missed. Participants' notes from meetings will need to be brought to the Review Session.

Review Session

There are three review activities:

1. Focusing on participants' changing perceptions of parents' knowledge, understanding and interest in their children's science.
2. Considering ways of involving parents in the assessment of science.
3. Drawing up a short-term policy re-creating a dialogue with parents.

The three review activities may be undertaken in one long session of up to 2 hours towards the end of a term or a school year, or after a period of reporting to parents. Alternatively, since several of the Stimulus and Classroom Activities are likely to generate a lot of evidence to review, you may prefer to deal with Review Activity 1 by having more regular, short sessions to look at this evidence.

1. Make sure that for each session you have some evidence from

participants to work on. This could include any of the following:

* the 'Diamond Sort' evidence from Stimulus Activity 1
* the Communication Chart for Stimulus Activity 2
* solutions to the simulation exercises in Stimulus Activity 3
* the draft communication from Stimulus Activity 4
* the monitoring evidence from Classroom Activity 1
* the notes from parents' meetings (Classroom Activity 2)

Arrange things so that evidence can be easily shared at the beginning of a session, either by displaying it or by having a short 'report-back' session to refresh memories. The purpose of these sessions is to focus on participants' changing perceptions of parents' knowledge, understanding and interest in their children's science; so you should make sure that you keep coming back to questions such as:

* What does this tell us?
* How reliable are our conclusions?
* Does it fit with other evidence?
* How good is our communication with parents now?

2 and 3. These are concluding activities, and are best undertaken together. They should bring together all the other evidence and ideas accumulated through the reviewing in Activity 1, to focus on policy for the short term (that is, the coming year, or the next round of assessment). You should aim at achieving some kind of 'Statement of Intent' from participants, to crystallize their ideas. You may want to combine this final review activity with that for Unit 5, 'Assessment'.

BRIEFING PAPER

How Well Informed Are Parents?

How much do the parents of children in your class know about the science your children are learning?

How do they value science, in relation to other school subjects?

How up-to-date and accurate is their information?

How sure are you about your answers to the above questions?

This paper starts with the above questions because, at the moment, there is a growing amount of evidence to show that parents' ideas about science in primary classrooms differ considerably from what actually happens or is beginning to happen. Obviously, the extent to which parents are well informed about developments such as National Curriculum science depends on many things. For example, our own research from 1990 onwards suggests that parents do not tend to ask teachers about their children's science, but they do pick up information from their children, who talk more about their science work than about other subjects. Parents who attend open evenings and science fairs at school tend to be much better informed. But as yet, many teachers do not seem to raise the question of science in a specific way at parents' evenings, and parents do not ask.

A useful starting point for any staff would be to try to establish what you already know about parents' knowledge and interest in science, and what your evidence is for this. Stimulus Activity 1 therefore tackles this question, and also examines the possible sources of information for parents about their children's science work and attainment. The point of this activity is to alert you to the danger of taking too many things for granted.

Other current research (Hughes *et al.* 1994) indicates that these same parents feel that teachers are doing a good job, and that they have a good relationship with the school. The main source of their information about the National Curriculum is their children's school; the vast majority want to know more about science; and most of them think that such information should come from the school, rather than from the LEA or the media. The DES, in its guide to the National Curriculum, has also stated clearly that it is the schools' responsibility to 'widen access to information about schools' curriculum plans and objectives – for individual pupils and classes in the case of parents', as well as to

> *give parents the information necessary to support an informed dialogue with the school and indeed with the children themselves about their achievements, progress and future work. . . . Sharing information and explaining its implications, rather than relying on snapshot information about grades or examination results, will encourage partnership between schools and parents.* (DES 1989)

So how can you, as a member of staff, adequately satisfy this demand in an area such as science where parents are probably less well informed than in any other? A starting point could be to review all the possible ways of communicating with parents, as described in Stimulus Activity 2.

Another important factor is to understand the specific questions to which parents tend to need to know answers.

What Do Parents Need to Know About?

Our evidence from the various pieces of research mentioned above is that parents need more information in the following areas:

1. *What science is in primary schools.* Many parents were never taught any science in their primary school days, and so their only notions of science come from secondary school. For example, many parents may see science as 'biology, physics and chemistry' and may be concerned about safety: parents often express anxiety about young children playing with bunsen burners and test-tubes, for example, even though most classrooms don't have a gas supply! Very few parents think of science as concerned with practical investigation and experimentation, which of course is a major element of the National Curriculum as well as a major dimension in these INSET Units.
2. *How children are taught science.* When asked if their children were taught science through topics or as a subject, fewer than 40 per cent of the parents were confident that they knew the answer. And amongst those who were sure, quite often school practice didn't match their beliefs. For example, parents of younger children tended to think they did topics, whilst parents of older children tended to think that they were mostly taught science as a subject, even though this was not often the case.
3. *When science should start for children.* Most parents by now are aware of the existence of the National Curriculum, but even recently only 62 per cent of the parents surveyed thought that their children should start doing science at 5+ or on entry into school, whilst 35 per cent didn't want science to begin until 7+, after Key Stage 1.
4. *What the school policy is on science.* This incorporates the concerns in the above questions, but it also expresses a wish to know the contents of the school's 'syllabus' for science, as many parents tended to see it. We have found that schools where parents are well informed often emphasize, in their communication at parents' meetings, what the children are going to do, rather than what they have done, as this is of more use to parents.
5. *How parents could help their children with science work at home.* No doubt parents who are used to helping their children with reading and maths are equally concerned, knowing that science is also a core subject, to do the same in science. Often, parents like to have advanced notice of topics, for example, so that they can provide their children with materials or even take them on appropriate visits.

Parents' Ideas about Assessment in Science

If parents are unclear about their children's science work, they are even less well informed about how it is assessed. What the research by Hughes *et al.* is showing is that only about half of the parents of Key Stage 1 children are happy about the assessment of 7-year-olds; the remainder have mixed feelings, and express many concerns, including concerns

about how it will be done, exams, labelling, the use of results, pressure on children, and so on. The first round of assessment in Key Stage 1 has cleared up some concerns about procedures; but there are many teething troubles and practicalities of the assessment process still to be sorted out. The whole question of the extent to which children will be tested in science by practical (rather than pencil-and-paper) tests is still in a state of uncertainty.

If parents are expressing such concerns, is it therefore appropriate to share with them the school's current state of knowledge, and the teachers' own concerns about assessment? Or is this, as some teachers think, 'unprofessional', as it carries an implicit criticism of the system and its creators? Will parents become even more concerned if they become aware of teachers' uncertainties? These are crucial questions for a school to address. Stimulus Activity 3 attempts to simulate some problems and questions which might arise, for you to address as a staff group.

A related question, which has drawn a wide range of responses from the teachers in schools in the South-West, is the extent to which parents can and should be involved in the assessment process. This will be addressed in Activity 6 below.

However, the fact that these uncertainties exist and need to be faced should not lead us into painting too gloomy a picture of parents' views and expectations. The most recent evidence concludes that:

1. the great majority of parents were happy with their choice of school, thought that the teachers were doing a good job, and were satisfied with their relationship with the school;
2. most parents approved in principle of a national curriculum, but were concerned about the restrictions and demands it made on teachers;
3. when asked what made a good school, parents mentioned relationships, staff, atmosphere and ethos much more frequently than academic results, resources or facilities;
4. most parents wanted to be better informed about current developments in their children's education. (Hughes *et al.* 1994)

Communicating about Science: Some Issues to Consider

Stimulus Activity 5 requires you to compare different ways of communicating with parents about a particular issue; namely, the reporting of children's attainment in relation to the two main components of science – investigation skills on the one hand, and knowledge on the other.

Some of the things you might need to consider are as follows.

1. Identifying investigation skills separately from knowledge. When teachers do report to parents about science, there is a tendency only to talk about the knowledge areas covered. Teachers are much less likely to report on children's skill development. And yet the National Curriculum requires that equal attention be given to investigation skills and knowledge in science.
2. What the levels mean. This might mean explaining the difference in skills between levels 1 and 5; it might mean trying to communicate an idea of 'three-ness', i.e. what level 3 involves in general.

3. The system of teacher assessment and SATs, and how each is to be carried out in your school. This might mean using examples of what your children have produced, to show the level at which they are working. You may need to look again at the various SEAC and SCAA documents relating to assessment and testing, to decide if any would be helpful to parents.
4. The record-keeping system you use, and its accessibility to parents, including an explanation of what the record means. Many schools have changed their way of reporting to parents in the last few years. Does your own report format adequately deal with the questions discussed above?
5. The Assessment Orders, and how much of them you communicate to parents.
6. Anticipating questions parents are likely to ask; e.g. about comparisons between the results of different children, comparisons between schools, how they know if their child is doing well/progressing in relation to other children, etc.
7. The difference between diagnostic and summative assessment.
8. What a SAT is, and how they are carried out in the classroom in science.

References

Department of Education and Science (DES) (1989) *From Policy to Practice*, London: HMSO.

Hughes, M., Wikeley, F. and Nash, T. (1994) *Parents and their Children's Schools*, Oxford: Blackwell.

Peacock, A. (1989) 'What parents think about science in primary schools', *Primary Science Review* 10.

Peacock, A. and Boulton, A. (1991) 'Parents' understanding of science in the National Curriculum', *Education* 3(13) (June).

STIMULUS ACTIVITIES

Activity 1

Below is a list of questions to which it might be useful to have answers, about parents' knowledge of their children's science work.

As a group, you should first put these roughly into order of priority by doing a 'Diamond Sort', as shown below:

Highest priority xxxxxxxxxx

xxxxxxxxxxxx xxxxxxxxxxxxx

xxxxxxxxxx xxxxxxxxxxxxx xxxxxxxxxxx

xxxxxxxxxxxx xxxxxxxxxxxxx

Lowest priority xxxxxxxxxxxx

Then begin with the highest-priority question, and discuss what you think *parents'* views are in relation to that question. For example, if the question chosen happened to be 'How do parents value their children doing science, in relation to work in language and maths?', you should share whatever evidence you have about parents' views on this matter, then try to decide:

* How much do we actually know about parents' views on this?
* Is our teaching of science based on any assumptions about parents' views on the value of science?
* How best could we find out more about this from parents?

When you have addressed these questions in relation to your top-priority question, try the same process in relation to the next priority. You may not get through all the priority questions; if not, you should decide how and when you are going to deal with the others, before the end of the session.

List of questions to which you might need answers

1. What do the parents of children in your class think science involves for their children?
2. How do they think you teach science to their children?
3. How important do they think science is? Do they value it as highly as English and maths? As highly as PE, music, art?
4. Do they think you use computers a lot in science?
5. Do they expect their children to plan and carry out investigations for themselves?
6. Do parents expect to see their children's science work?
7. Would the parents of the children in your class come to school for a science workshop if you organized one?
8. How do parents think you assess their children in science?
9. How many parents would help with science in your classroom if you invited them to?

Activity 2

The purpose of this activity is to try to work out all the ways of communicating with parents that are available, before making decisions about which is the most appropriate.

Communication is a two-way process, which is why we began by focusing on finding out. Often, we take communication to mean 'telling', when it is just as much about 'listening'.

Make a list of all the methods of communication with parents that you have ever used, and (if you are also a parent of school-age children) all the methods of communicating with teachers that you have used. You might try to organize your list into a table like this:

	As a teacher	*As a parent*
1. Ways of giving out information		
2. Ways of receiving information		
3. Ways of doing both simultaneously		

For example, writing letters comes into category 1; looking at a display of children's work comes into category 2; whilst having an interview with your child's class teacher should come into category 3.

Next, you should look at your list, and decide

* which of these methods do we as a staff use *most* frequently, and why?
* which of these methods do we use *least* frequently, and why?

This will probably get you talking about the advantages and drawbacks of different methods, which will be taken up in the Review Session. Round off your discussion by trying to re-think methods of communication; look at the possibility of trying strategies that you have not often tried, or look again at those which in the past have been unsuccessful.

Finally, consider an example of something that, at the moment, you need to communicate with parents about. Are there other ways which you haven't considered trying? If you find this difficult, Stimulus Activity 3 might help you get started.

Activity 3

Below are a number of letters and situations which are typical of those that crop up from time to time as a result of inadequate communication between school and home. All of these relate in some way or other to the science dimension of children's learning.

Sort out one or two which are most relevant to your own current position, and work out a practicable solution for each which would be acceptable to the majority of staff, without putting an undue burden on any particular individual.

Nick,

At the Governors' Meeting last week, we talked about the School Brochure and it was agreed that a few sections need up-dating for next year, including science. Can you draft me a piece, please, aimed at prospective parents, which deals specifically with new developments and shows how we are dealing with them, and how we teach science at present? Thanks.

Mrs W. particularly mentioned assessment, as you would have expected, so we'll have to be very clear on that, without frightening people. But we mustn't patronize our parents either; I have a feeling they are more tuned in to the NC than we sometimes imagine. You might also think about wording it in such a way that we encourage more parents to get involved with science in school.

The last point is about specialist teaching; let's talk about that first; tell me where you stand, given our need to make a couple of appointments in the not too distant future.

David Hinks

From Nick Barber (Science Coordinator)
To All Staff

Now that the new 'science guidelines' are in place and we have organized our schemes for the year, I'd like us to consider a

SCIENCE EVENING FOR PARENTS

some time later this term. *My* idea would be to talk to them briefly about the dreaded NC, and get all the jargon out of the way, then let them have a 'hands-on' session with some of the activities we are likely to be doing with children, particularly in the physical processes work (AT4).

But that's only my idea. Do you think we should do it? How would you want to operate the evening? Could you be there to 'man' (sorry!) an activity? Please let me have your arguments for and against, in a sealed brown envelope if necessary, and ideas for how to run the session. Which year(s) do you think we should start with?

Nick

Highfield
Cliff Road
Purbeton, Dorset

23rd March 1993

Dear Mr Hinks

I'm writing to you as Head because I'm somewhat frustrated by my son's reluctance to tell me anything about his work in science (or any subject, for that matter!) and am keen to know exactly what he is doing. When I was at school, it was Nature Study in the infants and juniors, then Biology, Physics and Chemistry up to 'O' level, but I realize things have changed. From talking to other parents, I get the feeling that some of the things which now come under science don't sound like science at all.

Mark has another year before he goes up to high school, and I would hope he is getting experience of all the main areas to prepare him for that. Can you put me in the picture? Would it help if I came in some time to talk to you about this?

Yours sincerely,

P. Smithson

Peter Smithson

Sue

Can you follow this up? Mark is one of yours, I think, for science.

Libby

Mrs Marsden rang to ask about Nicola's science work; she sounded a bit bothered about her using saws and things! Can you ring her and explain what science they do in year 1? Thanks!

Jane

15 Pennine Way
Swaledon,
N. Yorks.
16. 1. 93

Dear Mrs. Foster,

We know that Rachel is going to be doing tests in science later this year, and naturally we want her to do well, so we would be really grateful if you could suggest ways we could help her at home with her science. Is there a suitable book we could buy? We don't want to push her too hard she's only just gone 7 as you know, but we would like to help her prepare for the tests. Can you tell us how the tests will be done, as that might help us.

Yours sincerely
Mr. J. Parker

Activity 4

The purpose of this activity is to compare different modes of communication in practice.

The area of concern is *what children have learned in relation to the attainment targets in science.* The Briefing Paper sets out some of the things that you might have to bear in mind in its final section, 'Communicating about science; some issues to consider'.

Your group will focus on one of these modes of communication:

* a letter to parents
* a science fair or display of children's work
* a parents' open day/evening, or meeting
* using outside help (local media, higher education, sponsors, etc.)
* combinations of the above

Produce a draft or outline of what you would do to communicate with parents about the issue in question, explaining what it would involve for staff and for parents. Then draw up a balance sheet of plus points and minus points relating to the approach, considering such matters as cost, time, effectiveness, children's involvement, and so on.

CLASSROOM ACTIVITIES

Activity 1

The idea behind this activity is to carry out, as a group, an exercise in monitoring the effects of an actual communication with parents. The Stimulus Activities should have helped you to decide on your priorities in communicating with parents, and on the best way of ensuring good communication in this instance, remembering that communication is as much about listening as telling. At this stage, you should decide which communication you are going to monitor (for example, a report, parents' evening, a letter taken home by children, and so forth), and try to decide what you expect the outcome to be.

It would be helpful, next, to clarify exactly who will do what. In particular, you might break down the monitoring into separate activities, and allocate these to different people. These might mean keeping a systematic track on such things as the writing of the message; the timing of distribution; gathering responses; staff time involved in the process; making sense of responses; how many parents responded; what lessons were learned. It is primarily an attempt to be systematic about feedback to the group, to focus on the process of communication as well as the message.

Activity 2

Make a short list of headings referring to aspects of children's science you will raise with parents at the next round of parents' meetings. After the meetings, make a note of how much time you spent with each parent discussing science, and then analyse the way parents responded to this. Note down any questions parents asked about science, for use at the later Review Session.

REVIEW SESSION

Activity 1

Go back over what you have learned from each of the Classroom Activities, to see if it has altered your perception of parents' understanding. The evidence quoted in the Briefing Paper suggested that some teachers may have an out-of-date, stereotyped image of parents' knowledge and interest in science. What is the perception now amongst your group?

Activity 2

Consider each of the ways of involving parents in the assessment of science listed below, all of which have been used by teachers in recent years. In the light of your ideas from previous activities in this Unit, list the pluses and minuses of each. You might first report individual experiences of involving parents in science, or in assessment in other subject areas such as reading. You may also consider some of these ways inappropriate for parents to become involved; this should be a part of your discussion.

1. observing *individual* children with a specific Statement of Attainment in mind (e.g. observation/measurement skills);
2. observing a *group* to determine when they are ready to progress to a different activity/level;
3. working with groups not being assessed, in order to free the teacher to assess individuals;
4. managing science resources in order to give the teacher more time to concentrate on children's learning and diagnostic work;
5. working with the teacher to provide a 'second opinion' on children's attainment;
6. undertaking specific tasks at home with their children;
7. reporting on science activities their children do at home.

Activity 3

As a result of your discussions, is there staff consensus on links with parents concerning science? What does it cover?

Make a list of things you might realistically achieve during the coming year, in relation to:

1. *informing* parents about science;
2. *listening* to parents' questions about science;
3. *involving* parents in science.